Statistics in the Health Sciences

Series Editors
K. Dietz, M. Gail, K. Krickeberg, B. Singer

Statistics in the Health Sciences

Manton/Singer/Suzman: Forecasting the Health of Elderly Populations.

Salsburg: The Use of Restricted Significance Tests in Clinical Trials.

Kleinbaum: Logistic Regression: A Self-Learning Text.

David G. Kleinbaum

LOGISTIC REGRESSION

A Self-Learning Text

Springer-Verlag
New York   Berlin   Heidelberg   London   Paris
Tokyo   Hong Kong   Barcelona   Budapest

David G. Kleinbaum
Department of Epidemiology
Emory University
Atlanta, GA 30333
USA

Series Editors
K. Dietz
Institut für Medizinische Biometrie
Universität Tübingen
West Bahnhotstrasse 55
7400 Tübingen
Germany

M. Gail
National Cancer Institute
Rockville, MD 20892
USA

K. Krickeberg
3 Rue de L'Estrapade
75005 Paris
France

B. Singer
Department of Epidemiology
  and Public Health
Yale University
New Haven, CT 06510
USA

Library of Congress Cataloging-in-Publication Data
Kleinbaum, David G.
    Logistic regression: a self-learning text / David G. Kleinbaum.
      p.   cm.
    Includes bibliographical references and index.
    ISBN 0-387-94142-8
    1. Medicine--Research--Statistical methods. 2. Regression
analysis. 3. Logistic distribution.   I. Title.
    R853.S7K54 1992
    610'.72--dc20   93-27484

Printed on acid-free paper.

Production managed by Henry Krell; manufacturing supervised by Vincent Scelta.
Typeset by Impressions, Inc., Madison, WI.
Printed and bound by Edwards Brothers, Inc., Ann Arbor, MI.
Printed in the United States of America.

9 8 7 6 5 4 3 2 1

ISBN 0-387-94142-8 Springer-Verlag New York Berlin Heidelberg
ISBN 3-540-94142-8 Springer-Verlag Berlin Heidelberg New York

To my daughter,
Ruth Toby Kleinbaum

Preface

This text on logistic regression methods contains the following eight chapters:

1   Introduction to Logistic Regression
2   Important Special Cases of the Logistic Model
3   Computing the Odds Ratio in Logistic Regression
4   Maximum Likelihood Techniques: An Overview
5   Statistical Inferences Using Maximum Likelihood Techniques
6   Modeling Strategy Guidelines
7   Modeling Strategy for Assessing Interaction and Confounding
8   Analysis of Matched Data Using Logistic Regression

Each chapter contains a presentation of its topic in "lecture-book" format together with objectives, an outline, key formulae, practice exercises, and a test. The "lecture-book" has a sequence of illustrations and formulae in the left column of each page and a script in the right column. This format allows you to read the script in conjunction with the illustrations and formulae that high-light the main points, formulae, or examples being presented.

The reader may also purchase directly from the author audio-cassette tapes of each chapter. If you purchase the tapes, you may use the tape with the illustrations and formulae, ignoring the script. The use of the audiotape with the illustrations and formulae is intended to be similar to a lecture. An audio cassette player is the only equipment required. Tapes may be obtained by writing or calling the author at the following address: Department of Epidemiology, School of Public Health, Emory University, 1599 Clifton Rd. N.E., Atlanta, GA 30333, phone (404) 727–9667.

Suggestions for Use

This text is intended for self-study. The text may supplement material covered in a course, to review previously learned material or to provide the primary learning material in a self-instructional course or self-planned learning activity. A more individualized learning program may be particularly suitable to a working professional who does not have the time to participate in a regularly scheduled course. If you purchase the audiotape, you may consider using the tape without the script, although, if used this way, the learner will have difficulty visualizing any formulae or calculations being explained on the tape. Such use, however, may be beneficial for review purposes.

The order of the chapters represents what the author considers to be the logical order for learning about logistic regression. However, persons with some prior knowledge of the subject can choose whichever chapter appears appropriate to their learning needs, in whatever sequence desired.

In working with any chapter, the learner is encouraged first to read the abbreviated outline and the objectives and then work through the presentation, by using either the script with the picture sequence or the audiotape with the picture sequence. After finishing the presentation, the user is encouraged to read the detailed outline for a summary of the presentation, review the key formulae and other important information, work through the practice exercises, and, finally, complete the test to check what has been learned.

Recommended Preparation

The ideal preparation for this text is a course on quantitative methods in epidemiology and a course in applied multiple regression. The following are recommended references on these subjects, with suggested chapter readings:

Kleinbaum, D., Kupper, L., and Morgenstern, H., *Epidemiologic Research: Principles and Quantitative Methods*, Van Nostrand Reinhold Publishers, New York, 1982, Chaps. 1–19.

Kleinbaum, D., Kupper, L., and Muller, K., *Applied Regression Analysis and Other Multivariable Methods, Second Edition*, Duxbury Press, Boston, 1987, Chaps. 1–16.

A first course on the principles of epidemiologic research would be helpful as all modules in this series are written from the perspective of epidemiologic research. In particular, the learner should be familiar with the basic characteristics of epidemiologic study designs (follow-up, case control, and cross sectional) and should have some idea of the frequently encountered problem of controlling or adjusting for variables.

As for mathematics prerequisites, the reader should be familiar with natural logarithms and their relationship to exponentials (powers of e) and, more generally, should be able to read mathematical notation and formulae.

Acknowledgments

I wish to thank John Capelhorn at Sydney University's Department of Public Health for carefully reading the manuscript and providing meaningful critique and support. I also wish to thank the many EPID and BIOS students and colleagues at the University of North Carolina, particularly Holly Hill and Chris Nardo, for their positive feedback about the material in this text. I thank Don McNeil of Macquarie University and George Rubin of the New South Wales Department of Health for their encouragement and support. Finally, I wish to thank Bert Petersen of the Centers for Disease Control and Prevention for arranging institutional (i.e., CDC) support for the initial stages of this project and G. David Williamson, also at CDC and P, for support during the latter stages of this project.

Contents

Chapter 5   Statistical Inferences Using Maximum Likelihood Techniques   125

Chapter 6   Modeling Strategy Guidelines   161

Chapter 7   Modeling Strategy for Assessing Interaction and Confounding   191

1

Introduction to Logistic Regression

Introduction

This introduction to logistic regression describes the reasons for the popularity of the logistic model, the model form, how the model may be applied, and several of its key features, particularly how an odds ratio can be derived and computed for this model.

As preparation for this chapter, the reader should have some familiarity with the concept of a mathematical model, particularly a multiple-regression-type model involving independent variables and a dependent variable. Although knowledge of basic concepts of statistical inference is not required, the learner should be familiar with the distinction between population and sample, and the concept of a parameter and its estimate.

Abbreviated Outline

The outline below gives the user a preview of the material to be covered by the presentation. A detailed outline for review purposes follows the presentation.

Objectives   Upon completing this chapter, the learner should be able to:

1. Recognize the multivariable problem addressed by logistic regression in terms of the types of variables considered.
2. Identify properties of the logistic function that explain its popularity.
3. State the general formula for the logistic model and apply it to specific study situations.
4. Compute the estimated risk of disease development for a specified set of independent variables from a fitted logistic model.
5. Compute and interpret a risk ratio or odds ratio estimate from a fitted logistic model.
6. Identify the extent to which the logistic model is applicable to follow-up, case-control, and/or cross-sectional studies.
7. Identify the conditions required for estimating a risk ratio using a logistic model.
8. Identify the formula for the logit function and apply this formula to specific study situations.
9. Describe how the logit function is interpretable in terms of an "odds."
10. Interpret the parameters of the logistic model in terms of log odds.
11. Recognize that to obtain an odds ratio from a logistic model, you must specify **X** for two groups being compared.
12. Identify two formulae for the odds ratio obtained from a logistic model.
13. State the formula for the odds ratio in the special case of (0, 1) variables in a logistic model.
14. Describe how the odds ratio for (0, 1) variables is an "adjusted" odds ratio.
15. Compute the odds ratio, given an example involving a logistic model with (0, 1) variables and estimated parameters.
16. State a limitation regarding the types of variables in the model for use of the odds ratio formula for (0, 1) variables.

Presentation

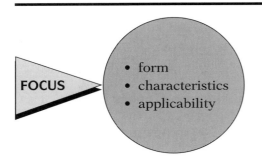

This presentation focuses on the basic features of logistic regression, a popular mathematical modeling procedure used in the analysis of epidemiologic data. We describe the **form** and key **characteristics** of the model. Also, we demonstrate the **applicability** of logistic modeling in epidemiologic research.

I. The Multivariable Problem

We begin by describing the multivariable problem frequently encountered in epidemiologic research. A typical question of researchers is: What is the relationship of one or more exposure (or study) variables (E) to a disease or illness outcome (D)?

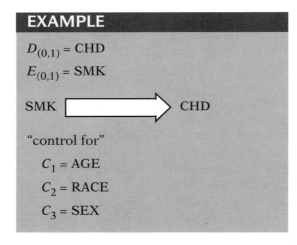

EXAMPLE

$D_{(0,1)}$ = CHD

$E_{(0,1)}$ = SMK

SMK ⟹ CHD

"control for"

C_1 = AGE

C_2 = RACE

C_3 = SEX

To illustrate, we will consider a dichotomous disease outcome with 0 representing **not diseased** and 1 representing **diseased.** The dichotomous disease outcome might be, for example, coronary heart disease (CHD) status, with subjects being classified as either 0 ("without CHD") or 1 ("with CHD").

Suppose, further, that we are interested in a single dichotomous exposure variable, for instance, smoking status, classified as "yes" or "no." The research question for this example is, therefore, to evaluate the extent to which smoking is associated with CHD status.

To evaluate the extent to which an exposure, like smoking, is associated with a disease, like CHD, we must often account or "control for" additional variables, such as age, race, and/or sex, which are not of primary interest. We have labeled these three control variables as C_1, C_2, and C_3.

In this example, the variable E (the exposure variable), together with C_1, C_2, and C_3 (the control variables), represent a collection of **independent** variables that we wish to use to describe or predict the **dependent** variable D.

Independent variables:
$$X_1, X_2, \ldots, X_k$$

X's may be E's, C's, or combinations

More generally, the independent variables can be denoted as X_1, X_2, and so on up to X_k where k is the number of variables being considered.

We have a **flexible** choice for the X's, which can represent any collection of exposure variables, control variables, or even combinations of such variables of interest.

EXAMPLE

$X_1 = E$	$X_4 = E \times C_1$
$X_2 = C_1$	$X_5 = C_1 \times C_2$
$X_3 = C_2$	$X_6 = E^2$

For example, we may have:

X_1 equal to an exposure variable E
X_2 and X_3 equal to control variables C_1 and C_2, respectively
X_4 equal to the product $E \times C_1$
X_5 equal to the product $C_1 \times C_2$
X_6 equal to E^2

The Multivariable Problem

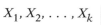
$$X_1, X_2, \ldots, X_k \qquad\longrightarrow\qquad D$$

The analysis:
  mathematical model

Logistic model:
  dichotomous D

Logistic is most popular

Whenever we wish to relate a set of X's to a dependent variable, like D, we are considering a **multivariable problem.** In the analysis of such a problem, some kind of **mathematical model** is typically used to deal with the complex interrelationships among many variables.

Logistic regression is a mathematical modeling approach that can be used to describe the relationship of several X's to a **dichotomous** dependent variable, such as D.

Other modeling approaches are possible also, but logistic regression is by far the most **popular** modeling procedure used to analyze epidemiologic data when the illness measure is dichotomous. We will show why this is true.

II. Why Is Logistic Regression Popular?

Logistic function:
$$f(z) = \frac{1}{1 + e^{-z}}$$

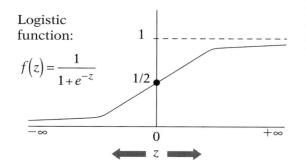

To explain the popularity of logistic regression, we show here the **logistic function,** which describes the mathematical form on which the **logistic model** is based. This function, called $f(z)$, is given by 1 over 1 plus e to the minus z. We have plotted the values of this function as z varies from $-\infty$ to $+\infty$.

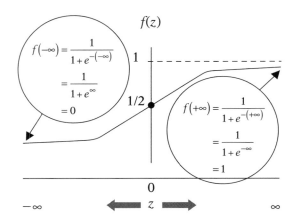

Notice, in the balloon on the left side of the graph, that when z is $-\infty$, the logistic function $f(z)$ equals 0.

On the right side, when z is $+\infty$, then $f(z)$ equals 1.

Range: $0 \le f(z) \le 1$

Thus, as the graph describes, the **range** of $f(z)$ is between 0 and 1, regardless of the value of z.

$0 \le$ probability ≤ 1
(individual risk)

The fact that the logistic function $f(z)$ **ranges between 0 and 1** is the primary reason the logistic model is so popular. The model is designed to describe a probability, which is always some number between 0 and 1. In epidemiologic terms, such a probability gives the **risk** of an individual getting a disease.

The **logistic model**, therefore, is set up to ensure that whatever estimate of risk we get, it will always be some number between 0 and 1. Thus, for the logistic model, we can never get a risk estimate either above 1 or below 0. This is not always true for other possible models, which is why the logistic model is often the first choice when a probability is to be estimated.

Shape:

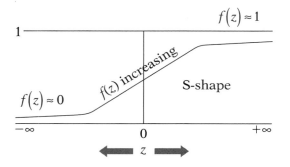

Another reason why the logistic model is popular derives from the **shape** of the logistic function. As shown in the graph, if we start at $z = -\infty$ and move to the right, then as z increases, the value of $f(z)$ hovers close to zero for a while, then starts to increase dramatically toward 1, and finally levels off around 1 as z increases toward $+\infty$. The result is an elongated, S-shaped picture.

z = index of combined risk factors

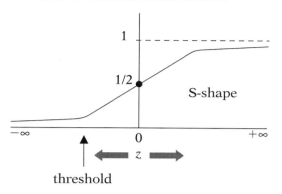

The S-shape of the logistic function appeals to epidemiologists if the variable z is viewed as representing an index that combines contributions of several risk factors, and $f(z)$ represents the risk for a given value of z.

Then, the S-shape of $f(z)$ indicates that the effect of z on an individual's risk is minimal for low z's until some **threshold** is reached. The risk then rises rapidly over a certain range of intermediate z values, and then remains extremely high around 1 once z gets large enough.

This **threshold** idea is thought by epidemiologists to apply to a variety of disease conditions. In other words, an S-shaped model is considered to be widely applicable for considering the multivariable nature of an epidemiologic research question.

SUMMARY

So, the logistic **model** is **popular** because the logistic **function**, on which the model is based, provides

- Estimates that must lie in the range between zero and one
- An appealing S-shaped description of the combined effect of several risk factors on the risk for a disease.

III. The Logistic Model

$$z = \alpha + \beta_1 X_1 + \beta_2 X_2 + \ldots + \beta_k X_k$$

Now, let's go from the logistic **function** to the **model**, which is our primary focus.

To obtain the logistic **model** from the logistic **function,** we write z as the linear sum α plus β_1 times X_1 plus β_2 times X_2, and so on to β_k times X_k, where the X's are independent variables of interest and α and the β_i are constant terms representing unknown parameters.

$$z = \alpha + \beta_1 X_1 + \beta_2 X_2 + \ldots + \beta_k X_k$$

In essence, then, z **is an index that combines the** X's.

$$f(z) = \frac{1}{1 + e^{-z}}$$
$$= \frac{1}{1 + e^{-(\alpha + \Sigma \beta_i X_i)}}$$

We now substitute the linear sum expression for z in the right-hand side of the formula for $f(z)$ to get the expression $f(z)$ equals 1 over 1 plus e to minus the quantity α plus the sum of $\beta_i X_i$ for i ranging from 1 to k. Actually, to view this expression as a mathematical model, we must place it in an epidemiologic context.

Epidemiologic framework

$X_1, X_2, \ldots, X_k$ measured at T_0

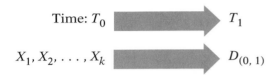

Time: T_0 T_1

$X_1, X_2, \ldots, X_k$ $D_{(0,\, 1)}$

$P(D=1 | X_1, X_2, \ldots, X_k)$

DEFINITION
Logistic model:

$P(D=1 | X_1, X_2, \ldots, X_k)$

$$= \frac{1}{1 + e^{-(\alpha + \Sigma \beta_i X_i)}}$$

$\uparrow \quad \uparrow$

unknown parameters

NOTATION
$P(D=1 | X_1, X_2, \ldots, X_k)$

$= P(\mathbf{X})$

Model formula:

$$P(\mathbf{X}) = \frac{1}{1 + e^{-(\alpha + \Sigma \beta_i X_i)}}$$

The logistic model considers the following general **epidemiologic study framework:** We have observed independent variables X_1, X_2, and so on up to X_k on a group of subjects, for whom we have also determined disease status, as either 1 if "with disease" or 0 if "without disease."

We wish to use this information to describe the probability that the disease will develop during a defined study period, say T_0 to T_1, in a disease-free individual with independent variable values X_1, X_2, up to X_k which are measured at T_0.

The probability being modeled can be denoted by the conditional probability statement $P(D=1 \mid X_1, X_2, \ldots, X_k)$.

The model is defined as **logistic** if the expression for the probability of developing the disease, given the X's, is 1 over 1 plus e to minus the quantity α plus the sum from i equals 1 to k of β_i times X_i.

The terms α and β_i in this model represent **unknown parameters** that we need to estimate based on data obtained on the X's and on D (disease outcome) for a group of subjects.

Thus, if we knew the parameters α and the β_i and we had determined the values of X_1 through X_k for a particular disease-free individual, we could use this formula to plug in these values and obtain the probability that this individual would develop the disease over some defined follow-up time interval.

For notational convenience, we will denote the probability statement $P(D=1 \mid X_1, X_2, \ldots, X_k)$ as simply $P(\mathbf{X})$ where the **bold X** is a shortcut notation for the collection of variables X_1 through X_k.

Thus, the logistic model may be written as $P(\mathbf{X})$ equals 1 over 1 plus e to minus the quantity α plus the sum $\beta_i X_i$.

IV. Applying the Logistic Model Formula

EXAMPLE

$D = \text{CHD}_{(0,\,1)}$

$X_1 = \text{CAT}_{(0,\,1)}$

$X_2 = \text{AGE}_{\text{continuous}}$

$X_3 = \text{ECG}_{(0,\,1)}$

$n = 609$ white males

9-year follow-up

$$P(\mathbf{X}) = \frac{1}{1 + e^{-(\alpha + \beta_1 \text{CAT} + \beta_2 \text{AGE} + \beta_3 \text{ECG})}}$$

DEFINITION
fit: use data to estimate

$\alpha, \beta_1, \beta_2, \beta_3$

NOTATION
hat= ˆ

$\alpha \ \beta_1 \ \beta_2 \ / \ \hat{\alpha} \ \hat{\beta}_1 \ \hat{\beta}_2$

Method of estimation:
   maximum likelihood (ML)—
   see Chapters 4 and 5

EXAMPLE

$\hat{\alpha} = -3.911$

$\hat{\beta}_1 = 0.652$

$\hat{\beta}_2 = 0.029$

$\hat{\beta}_3 = 0.342$

To illustrate the use of the logistic model, suppose the disease of interest is D equals CHD. Here CHD is 1 if a person has the disease and 0 if not.

We have three independent variables of interest: X_1=CAT, X_2=AGE, and X_3=ECG. CAT stands for catecholamine level and is 1 if high and 0 if low, AGE is continuous, and ECG denotes electrocardiogram status and is 1 if abnormal and 0 if normal.

We have a data set of 609 white males on which we measured CAT, AGE, and ECG at the start of study. These people were then followed for 9 years to determine CHD status.

Suppose that in the analysis of this data set, we consider a logistic model given by the expression shown here.

We would like to "**fit**" this model; that is, we wish to use the data set to estimate the unknown parameters α, β_1, β_2, and β_3.

Using common statistical notation, we distinguish the parameters from their estimators by putting a **hat** symbol on top of a parameter to denote its estimator. Thus, the estimators of interest here are α "hat," β_1 "hat," β_2 "hat," and β_3 "hat."

The method used to obtain these estimates is called **maximum likelihood** (ML). In two later chapters (Chapters 4 and 5), we describe how the ML method works and how to test hypotheses and derive confidence intervals about model parameters.

Suppose the results of our model fitting yield the following estimated parameters:

α "hat" = -3.911
β_1 "hat" = 0.652
β_2 "hat" = 0.029
β_3 "hat" = 0.342

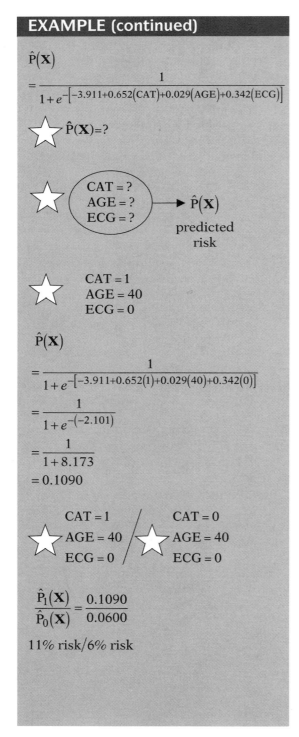

$$\hat{P}(\mathbf{X})$$

$$= \frac{1}{1+e^{-\left[-3.911+0.652(\mathrm{CAT})+0.029(\mathrm{AGE})+0.342(\mathrm{ECG})\right]}}$$

$$\hat{P}(\mathbf{X})=?$$

CAT = ?
AGE = ?
ECG = ? → $\hat{P}(\mathbf{X})$
predicted risk

CAT = 1
AGE = 40
ECG = 0

$$\hat{P}(\mathbf{X})$$

$$= \frac{1}{1+e^{-\left[-3.911+0.652(1)+0.029(40)+0.342(0)\right]}}$$

$$= \frac{1}{1+e^{-(-2.101)}}$$

$$= \frac{1}{1+8.173}$$

$$= 0.1090$$

CAT = 1 / CAT = 0
AGE = 40 / AGE = 40
ECG = 0 / ECG = 0

$$\frac{\hat{P}_1(\mathbf{X})}{\hat{P}_0(\mathbf{X})} = \frac{0.1090}{0.0600}$$

11% risk/6% risk

Our fitted model thus becomes P "hat" of **X** equals 1 over 1 plus e to minus the linear sum -3.911 plus 0.652 times CAT plus 0.029 times AGE plus 0.342 times ECG. We have replaced P by P "hat" on the left-hand side of the formula because our estimated model will give us an estimated probability, not the exact probability.

Suppose we want to use our fitted model, to obtain the predicted risk for a **certain individual.**

To do so, we would need to specify the values of the independent variables (CAT, AGE, ECG) for this individual, and then plug these values into the formula for the fitted model to compute the estimated probability, P "hat," for this individual. This estimate is often called a "predicted risk," or simply "risk."

To illustrate the calculation of a predicted risk, suppose we consider an individual with CAT=1, AGE=40, and ECG=0.

Plugging these values into the fitted model gives us 1 over 1 plus e to minus the quantity -3.911 plus 0.652 times 1 plus 0.029 times 40 plus 0.342 times 0. This expression simplifies to 1 over 1 plus e to minus the quantity -2.101, which further reduces to 1 over 1 plus 8.173, which yields the value **0.1090**.

Thus, for a person with CAT=1, AGE=40, and ECG=0, the predicted risk obtained from the fitted model is 0.1090. That is, this person's estimated risk is about 11%.

Here, for the same fitted model, we compare the predicted risk of a person with CAT=1, AGE=40, and ECG=0 with that of a person with CAT=0, AGE=40, and ECG=0.

We previously computed the risk value of 0.1090 for the first person. The second probability is computed the same way, but this time we must replace CAT=1 with CAT=0. The predicted risk for this person turns out to be **0.0600**. Thus, using the fitted model, the person with a high catecholamine level has an **11% risk** for CHD, whereas the person with a low catecholamine level has a **6% risk** for CHD over the period of follow-up of the study.

EXAMPLE

$$\frac{\hat{P}_1(X)}{\hat{P}_0(X)} = \frac{0.109}{0.060} = 1.82 \quad \text{risk ratio (RR)}$$

Note that, in this example, if we divide the predicted risk of the person with high catecholamine by that of the person with low catecholamine, we get a **risk ratio** estimate, denoted by **RR,** of **1.82.** Thus, using the fitted model, we find that the person with high CAT has almost twice the risk of the person with low CAT, assuming both persons are of AGE 40 and have no previous ECG abnormality.

- RR (direct method)

We have just seen that it is possible to use a logistic model to obtain a risk ratio estimate that compares two types of individuals. We will refer to the approach we have illustrated above as the **direct method** for estimating RR.

Conditions for RR (direct method)
  ✓ follow-up study
  ✓ specify all X's

Two conditions must be satisfied to estimate RR directly. First, we must have a **follow-up study** so that we can legitimately estimate individual risk. Second, for the two individuals being compared, we must **specify values for all the independent variables** in our fitted model to compute risk estimates for each individual.

If either of the above conditions is not satisfied, then we cannot estimate RR directly. That is, if our study design is not a follow-up study *or* if some of the X's are not specified, we cannot estimate RR directly. Nevertheless, it may be possible to estimate RR **indirectly.** To do this, we must first compute an **odds ratio,** usually denoted as **OR,** and we must make some assumptions that we will describe shortly.

- RR (indirect method)
  ✓ OR
  ✓ assumptions

- OR: direct estimate from
  ✓ follow-up
  ✓ case control
  ✓ cross sectional

In fact, **the odds ratio (OR),** not the risk ratio (RR), *is the only measure of association* **directly estimated** *from a logistic model (without requiring special assumptions), regardless of whether the study design is* **follow-up, case control,** *or* **cross sectional.** To see how we can use the logistic model to get an odds ratio, we need to look more closely at some of the features of the model.

V. Study Design Issues

An important feature of the logistic model is that it is defined with a **follow-up study orientation.** That is, as defined, this model describes the probability of developing a disease of interest expressed as a function of independent variables presumed to have been measured at the start of a fixed follow-up period. For this reason, it is natural to wonder whether the model can be applied to case-control or cross-sectional studies.

★ Follow-up study orientation

$$X_1, X_2, \ldots, X_k \quad \longrightarrow \quad D_{(0, 1)}$$

✓ **case control**
✓ **cross sectional**

Breslow and Day (1981)
Prentice and Pike (1979)

    robust conditions
    case-control studies

    robust conditions
    cross-sectional studies

Case control:

D  E

Follow-up:

E  D

Treat case control like follow-up

LIMITATION

case-control and
cross-sectional studies:

✓ OR

The answer is **yes:** logistic regression can be applied to study designs other than follow-up.

Two papers, one by **Breslow and Day** in 1981 and the other by **Prentice and Pike** in 1979 have identified certain **"robust" conditions** under which the logistic model can be used with case-control data. "Robust" means that the conditions required, which are quite complex mathematically and equally as complex to verify empirically, apply to a large number of data situations that actually occur.

The reasoning provided in these papers carries over to **cross-sectional studies** also, though this has not been explicitly demonstrated in the literature.

In terms of **case-control** studies, it has been shown that even though cases and controls are selected first, after which previous exposure status is determined, the analysis may proceed as if the selection process were the other way around, as in a follow-up study.

In other words, even with a case-control design, one can pretend, when doing the analysis, that the dependent variable is disease outcome and the independent variables are exposure status plus any covariates of interest. When using a logistic model with a case-control design, you can treat the data as if it came from a follow-up study, and still get a *valid* answer.

Although logistic modeling is applicable to case-control and cross-sectional studies, there is one important **limitation** in the analysis of such studies. Whereas in follow-up studies, as we demonstrated earlier, a fitted logistic model can be used to predict the risk for an individual with specified independent variables, this model cannot be used to predict individual risk for case-control or cross-sectional studies. In fact, *only estimates of* **odds ratios** *can be obtained for case-control and cross-sectional studies.*

Simple Analysis

	$E = 1$	$E = 0$
$D = 1$	a	b
$D = 0$	c	d

Risk: only in follow-up
OR: case-control or cross-sectional

$$\widehat{OR} = ad/bc$$

Case-control and cross-sectional studies:

$$= \frac{\hat{P}(E=1 \mid D=1) \Big/ \hat{P}(E=0 \mid D=1)}{\hat{P}(E=1 \mid D=0) \Big/ \hat{P}(E=0 \mid D=0)}$$

$$\left.\begin{array}{l} \hat{P}(E=1 \mid D=1) \\ \hat{P}(E=1 \mid D=0) \end{array}\right\} P(E \mid D)$$

Risk: $P(D|E)$
$\downarrow$

$$RR = \frac{\hat{P}(D=1 \mid E=1)}{\hat{P}(D=1 \mid E=0)}$$

The fact that only odds ratios, not individual risks, can be estimated from logistic modeling in case-control or cross-sectional studies is not surprising. This phenomenon is a carryover of a principle applied to simpler data analysis situations, in particular, to the simple analysis of a 2×2 table, as shown here.

For a 2×2 table, **risk estimates** can be used *only* if the data derive from a follow-up study, whereas only **odds ratios** are appropriate if the data derive from a case-control or cross-sectional study.

To explain this further, recall that for 2×2 tables, the odds ratio is calculated as OR "hat" equals a times d over b times c, where a, b, c, and d are the cell frequencies inside the table.

In case-control and cross-sectional studies, this OR formula can alternatively be written, as shown here, as a ratio involving probabilities for exposure status conditional on disease status.

In this formula, for example, the term $\hat{P}(E{=}1 \mid D{=}1)$ is the estimated probability of being exposed, given that you are diseased. Similarly, the expression $\hat{P}(E{=}1 \mid D{=}0)$ is the estimated probability of being exposed given that you are not diseased. All the probabilities in this expression are of the general form $P(E \mid D)$.

In contrast, in follow-up studies, formulae for risk estimates are of the form $P(D \mid E)$, in which the exposure and disease variables have been switched to the opposite side of the "given" sign.

For example, the risk ratio formula for follow-up studies is shown here. Both the numerator and denominator in this expression are of the form $P(D \mid E)$.

Case-control or cross-sectional studies:

~~P(D|E)~~

✓ P(E|D)

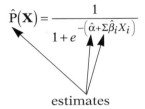

$$\hat{P}(\mathbf{X}) = \frac{1}{1 + e^{-\left(\hat{\alpha} + \Sigma\hat{\beta}_i X_i\right)}}$$

estimates

Case control:

$\cancel{\hat{\alpha}} \Rightarrow \hat{P}(\cancel{\mathbf{X}})$

Follow-up:

$\hat{\alpha} \Rightarrow \hat{P}(\mathbf{X})$

Case-control and cross-sectional:

✓ $\hat{\beta}_i$, $\widehat{OR}$

Thus, in case-control or cross-sectional studies, risk estimates cannot be estimated because such estimates require conditional probabilities of the form $P(D|E)$, whereas only estimates of the form $P(E|D)$ are possible. This classic feature of a simple analysis also carries over to a logistic analysis.

There is a simple **mathematical explanation** for why predicted risks cannot be estimated using logistic regression for case-control studies. To see this, we consider the parameters α and the β's in the logistic model. To get a predicted risk P "hat" of **X** from fitting this model, we must obtain valid estimates of α and the β's, these estimates being denoted by "hats" over the parameters in the mathematical formula for the model.

When using logistic regression for case-control data, the parameter α cannot be validly estimated without knowing the sampling fraction of the population. Without having a "good" estimate of α, we cannot obtain a good estimate of the predicted risk P "hat" of **X** because α "hat" is required for the computation.

In contrast, in follow-up studies, α can be estimated validly, and, thus, $P(\mathbf{X})$ can also be estimated.

Now, although α cannot be estimated from a case-control or cross-sectional study, the β's can be estimated from such studies. As we shall see shortly, the β's provide information about odds ratios of interest. Thus, even though we cannot estimate α in such studies, and therefore cannot obtain predicted risks, we can, nevertheless, obtain estimated measures of association in terms of odds ratios.

Note that if a logistic model is fit to case-control data, most computer packages carrying out this task will provide numbers corresponding to all parameters involved in the model, including α. This is illustrated here with some fictitious numbers involving three variables, X_1, X_2, and X_3. These numbers include a value corresponding to α, namely, -4.5, which corresponds to the constant on the list.

EXAMPLE

Printout

Variable	Coefficient
constant	$-4.50 = \hat{\alpha}$
X_1	$0.70 = \hat{\beta}_1$
X_2	$0.05 = \hat{\beta}_2$
X_3	$0.42 = \hat{\beta}_3$

$\cancel{\alpha}$

EXAMPLE (repeated)

Printout

Variable	Coefficient
constant	$-4.50 = \hat{\alpha}$
X_1	$0.70 = \hat{\beta}_1$
X_2	$0.05 = \hat{\beta}_2$
X_3	$0.42 = \hat{\beta}_3$

However, according to mathematical theory, the value provided for the constant does not really estimate α. In fact, this value estimates some other parameter of no real interest. Therefore, an investigator should be forewarned that, even though the computer will print out a number corresponding to the constant α, the number will not be an appropriate estimate of α in case-control or cross-sectional studies.

SUMMARY

	Logistic Model	$\hat{P}(\mathbf{X})$	OR
Follow-up	✓	✓	✓
Case-control	✓	X	✓
Cross-sectional	✓	X	✓

We have thus described that the logistic model can be applied to case-control and cross-sectional data, even though it is intended for a follow-up design. When using case-control or cross-sectional data, however, a key limitation is that you cannot estimate risks like P "hat" of **X**, even though you can still obtain odds ratios. This limitation is not extremely severe if the goal of the study is to obtain a valid estimate of an exposure–disease association in terms of an odds ratio.

VI. Risk Ratios Versus Odds Ratios

OR

vs. **?** follow-up study

RR

The use of an odds ratio estimate may still be of some concern, particularly when the study is a follow-up study. In follow-up studies, it is commonly preferred to estimate a risk ratio rather than an odds ratio.

EXAMPLE

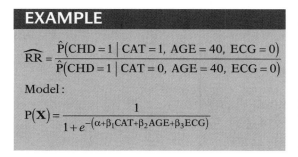

$$\widehat{RR} = \frac{\hat{P}(CHD = 1 \mid CAT = 1, \ AGE = 40, \ ECG = 0)}{\hat{P}(CHD = 1 \mid CAT = 0, \ AGE = 40, \ ECG = 0)}$$

Model :

$$P(\mathbf{X}) = \frac{1}{1 + e^{-(\alpha + \beta_1 CAT + \beta_2 AGE + \beta_3 ECG)}}$$

We previously illustrated that a risk ratio can be estimated for follow-up data provided all the independent variables in the fitted model are specified. In the example, we showed that we could estimate the risk ratio for CHD by comparing high catecholamine persons (that is, those with CAT=1) to low catecholamine persons (those with CAT=0), given that both persons were 40 years old and had no previous ECG abnormality. Here, we have specified values for all the independent variables in our model, namely, CAT, AGE, and ECG, for the two types of persons we are comparing.

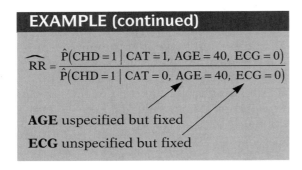

EXAMPLE (continued)

$$\widehat{RR} = \frac{\hat{P}(CHD = 1 \mid CAT = 1, \ AGE = 40, \ ECG = 0)}{\hat{P}(CHD = 1 \mid CAT = 0, \ AGE = 40, \ ECG = 0)}$$

AGE uspecified but fixed

ECG unspecified but fixed

Nevertheless, it is more common to obtain an estimate of a risk ratio or odds ratio without explicitly specifying the control variables. In our example, for instance, it is typical to compare high CAT with low CAT persons keeping the control variables like AGE and ECG fixed but unspecified. In other words, the question is typically asked, What is the effect of the CAT variable controlling for AGE and ECG, considering persons who have the same AGE and ECG *regardless* of the values of these two variables?

Control variables unspecified:

$\widehat{OR}$ directly

$\widehat{RR}$ indirectly
provided $\widehat{OR} \approx \widehat{RR}$

When the control variables are generally considered to be fixed, but **unspecified,** as in the last example, we can use logistic regression to obtain an estimate of the odds ratio **directly,** but we cannot estimate the risk ratio. We can, however, stretch our interpretation to obtain a risk ratio **indirectly** provided we are willing to make certain assumptions. The key **assumption** here is that the odds ratio provides a good approximation to the risk ratio.

$\widehat{OR} \approx \widehat{RR}$ if rare disease

From previous exposure to epidemiologic principles, you may recall that one way to justify an odds ratio approximation for a risk ratio is to assume that the disease is rare. Thus, if we invoke the **rare disease assumption,** we can assume that the odds ratio estimate from a logistic regression model approximates a risk ratio.

Rare disease		OR	RR
	yes	✓	✓
	no	✓	?

If we cannot invoke the rare disease assumption, we cannot readily claim that the odds ratio estimate obtained from logistic modeling approximates a risk ratio. The investigator, in this case, may have to review the specific characteristics of the study before making a decision. It may be necessary to conclude that the odds ratio is a satisfactory measure of association in its own right for the current study.

VII. Logit Transformation

OR: Derive and Compute

Having described why the odds ratio is the primary parameter estimated when fitting a logistic regression model, we now explain how an odds ratio is derived and computed from the logistic model.

Logit

$$\text{logit } P(\mathbf{X}) = \ln_e\left[\frac{P(\mathbf{X})}{1 - P(\mathbf{X})}\right]$$

where

$$P(\mathbf{X}) = \frac{1}{1 + e^{-(\alpha + \Sigma\beta_i X_i)}}$$

(1) $P(\mathbf{X})$

(2) $1 - P(\mathbf{X})$

(3) $\dfrac{P(\mathbf{X})}{1 - P(\mathbf{X})}$

(4) $\ln_e\left[\dfrac{P(\mathbf{X})}{1 - P(\mathbf{X})}\right]$

EXAMPLE

(1) $P(\mathbf{X}) = 0.110$

(2) $1 - P(\mathbf{X}) = 0.890$

(3) $\dfrac{P(\mathbf{X})}{1 - P(\mathbf{X})} = \dfrac{0.110}{0.890} = 0.123$

(4) $\ln_e\left[\dfrac{P(\mathbf{X})}{1 - P(\mathbf{X})}\right] = \ln(0.123) = -2.096$

i.e., logit $(0.110) = -2.096$

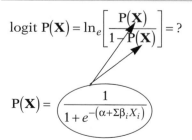

$$\text{logit } P(\mathbf{X}) = \ln_e\left[\frac{P(\mathbf{X})}{1 - P(\mathbf{X})}\right] = ?$$

$$P(\mathbf{X}) = \frac{1}{1 + e^{-(\alpha + \Sigma\beta_i X_i)}}$$

To begin the description of the odds ratio in logistic regression, we present an alternative way to write the logistic model, called the **logit form** of the model. To get the **logit** from the logistic model, we make a transformation of the model.

The **logit transformation,** denoted as **logit** P(**X**), is given by the natural log (i.e., to the base e) of the quantity P(**X**) divided by one minus P(**X**), where P(**X**) denotes the logistic model as previously defined.

This transformation allows us to compute a number, called **logit** P(**X**), for an individual with independent variables given by **X**. We do so by:

(1) computing P(**X**) and
(2) 1 minus P(**X**) separately, then
(3) dividing one by the other, and finally
(4) taking the natural log of the ratio.

For example, if P(**X**) is 0.110, then

1 minus P(**X**) is 0.890,

the ratio of the two quantities is 0.123,

and the log of the ratio is −2.096.

That is, the **logit** of 0.110 is −2.096.

Now we might ask, **what general formula do we get when we plug the logistic model form into the logit function? What kind of interpretation can we give to this formula? How does this relate to an odds ratio?**

Let us consider the formula for the logit function. We start with P(**X**), which is 1 over 1 plus e to minus the quantity α plus the sum of the $\beta_i X_i$.

$$1 - P(\mathbf{X}) = 1 - \frac{1}{1 + e^{-(\alpha + \Sigma\beta_i X_i)}}$$

put over common denominator

$$= \frac{e^{-(\alpha + \Sigma\beta_i X_i)}}{1 + e^{-(\alpha + \Sigma\beta_i X_i)}}$$

$$\frac{P(\mathbf{X})}{1 - P(\mathbf{X})} = \frac{\dfrac{1}{1 + e^{-(\alpha + \Sigma\beta_i X_i)}}}{\dfrac{e^{-(\alpha + \Sigma\beta_i X_i)}}{1 + e^{-(\alpha + \Sigma\beta_i X_i)}}}$$

$$= e^{(\alpha + \Sigma\beta_i X_i)}$$

$$\ln_e\left[\frac{P(\mathbf{X})}{1 - P(\mathbf{X})}\right] = \ln_e\left[e^{(\alpha + \Sigma\beta_i X_i)}\right]$$

$$= \underbrace{(\alpha + \Sigma\beta_i X_i)}_{\text{linear sum}}$$

Logit form:

$$\boxed{\begin{array}{l} \text{logit } P(\mathbf{X}) = \alpha + \Sigma\beta_i X_i \\[2mm] \text{where} \\[2mm] P(\mathbf{X}) = \dfrac{1}{1 + e^{-(\alpha + \Sigma\beta_i X_i)}} \end{array}}$$

logit $P(\mathbf{X})$ ➡ **?**   OR

$$\frac{P(\mathbf{X})}{1 - P(\mathbf{X})} = \text{odds for individual } X$$

$$\text{odds} = \frac{P}{1 - P}$$

Also, using some algebra, we can write $1 - P(\mathbf{X})$ as:

e to minus the quantity α plus the sum of $\beta_i X_i$ divided by one over 1 plus e to minus α plus the sum of the $\beta_i X_i$.

If we divide $P(\mathbf{X})$ by $1 - P(\mathbf{X})$, then the denominators cancel out,

and we obtain e to the quantity α plus the sum of the $\beta_i X_i$.

We then compute the natural log of the formula just derived to obtain:

the linear sum α plus the sum of $\beta_i X_i$.

Thus, the **logit** of $P(\mathbf{X})$ simplifies to the **linear sum** found in the denominator of the formula for $P(\mathbf{X})$.

For the sake of convenience, many authors describe the logistic model in its logit form rather than in its original form as $P(\mathbf{X})$. Thus, when someone describes a model as **logit** $P(\mathbf{X})$ equal to a linear sum, we should recognize that a logistic model is being used.

Now, having defined and expressed the formula for the logit form of the logistic model, we ask, **where does the odds ratio come in?** As a preliminary step to answering this question, we first look more closely at the definition of the logit function. In particular, the quantity $P(\mathbf{X})$ divided by $1 - P(\mathbf{X})$, whose log value gives the **logit,** describes the **odds** for developing the disease for a person with independent variables specified by $\mathbf{X}$.

In its simplest form, an **odds** is the ratio of the probability that some event will occur over the probability that the same event will not occur. The formula for an odds is, therefore, of the form P divided by $1 - P$, where P denotes the probability of the event of interest.

EXAMPLE

$P = 0.25$

$$\text{odds} = \frac{P}{1-P} = \frac{0.25}{0.75} = \frac{1}{3}$$

$\dfrac{1}{3}$ $\leftarrow$ event occurs
$\leftarrow$ event does not occur

3 to 1 event will not happen

For example, if P equals 0.25, then $1-P$, the probability of the opposite event, is 0.75 and the **odds** is 0.25 over 0.75, or one-third.

An **odds** of one-third can be interpreted to mean that the probability of the event occurring is one-third the probability of the event not occurring. Alternatively, we can state that the **odds** are **3 to 1** that the event will not happen.

$$\text{odds} : \left[\frac{P(\mathbf{X})}{1-P(\mathbf{X})}\right] \text{vs.} \frac{P}{1-P}$$

describes risk in logistic model for individual $\mathbf{X}$

The expression $P(\mathbf{X})$ divided by $1-P(\mathbf{X})$ has essentially the same interpretation as P over $1-P$, which ignores $\mathbf{X}$.

The main difference between the two formulae is that the expression with the $\mathbf{X}$ is more specific. That is, the formula with $\mathbf{X}$ assumes that the probabilities describe the risk for developing a disease, that this risk is determined by a logistic model involving independent variables summarized by $\mathbf{X}$, and that we are interested in the odds associated with a particular specification of $\mathbf{X}$.

$$\text{logit } P(\mathbf{X}) = \ln_e\left[\frac{P(\mathbf{X})}{1-P(\mathbf{X})}\right]$$

$$= \text{log odds for individual } \mathbf{X}$$

$$= \alpha + \Sigma\beta_i X_i$$

Thus, the logit form of the logistic model, shown again here, gives an expression for the **log odds** of developing the disease for an individual with a specific set of X's.

And, mathematically, this expression equals α plus the sum of the $\beta_i X_i$.

EXAMPLE

all $X_i = 0$: logit $P(\mathbf{X}) = ?$

$$\text{logit } P(\mathbf{X}) = \alpha + \Sigma\beta_i X_i$$

$$\text{logit } P(\mathbf{X}) \Rightarrow \alpha$$

(1) $\alpha = $ log odds for individual with all
$X_i = 0$

As a simple example, consider what the **logit** becomes when all the X's are 0. To compute this, we need to work with the mathematical formula, which involves the unknown parameters and the X's.

If we plug in 0 for all the X's in the formula, we find that the logit of $P(\mathbf{X})$ reduces simply to α.

Because we have already seen that any logit can be described in terms of an **odds,** we can interpret this result to give some meaning to the parameter α.

One interpretation is that α gives the **log odds** for a person with zero values for all X's.

EXAMPLE (continued)

(2) α = log of background odds

LIMITATION

All $X_i = 0$ for any individual?

↓

AGE $\neq 0$

WEIGHT $\neq 0$

(2) α = log of background odds

DEFINITION

background odds: ignores all X's

model : $P(\mathbf{X}) = \dfrac{1}{1 + e^{-\alpha}}$

α ✓

β_i?

$X_1, X_2, \ldots, X_i, \ldots, X_k$

fixed   varies   fixed

EXAMPLE

CAT varies from 0 to 1;

AGE = 40, ECG = 0

⎱⎰

fixed

logit $P(\mathbf{X}) = \alpha + \beta_1 CAT + \beta_2 AGE + \beta_3 ECG$

A second interpretation is that α gives the **log** of the **background**, *or* **baseline, odds.**

The first interpretation for α, which considers it as the **log odds** for a person with 0 values for all X's, has a serious limitation: There may not be any person in the population of interest with zero values on all the X's.

For example, no subject could have zero values for naturally occurring variables, like age or weight. Thus, it would not make sense to talk of a person with zero values for all X's.

The second interpretation for α is more appealing: to describe it as the **log** of the **background, *or* baseline, odds.**

By background odds, we mean the odds that would result for a logistic model without any X's at all.

The form of such a model is 1 over 1 plus e to minus α. We might be interested in this model to obtain a baseline risk or odds estimate that ignores all possible predictor variables. Such an estimate can serve as a starting point for comparing other estimates of risk or odds when one or more X's are considered.

Because we have given an interpretation to α, can we also give an interpretation to β_i? Yes, we can, in terms of either **odds** or **odds ratios.** We will turn to odds ratios shortly.

With regard to the odds, we need to consider what happens to the logit when only one of the X's varies while keeping the others fixed.

For example, if our X's are CAT, AGE, and ECG, we might ask what happens to the logit when CAT changes from 0 to 1, given an AGE of 40 and an ECG of 0.

To answer this question, we write the model in **logit form** as $\alpha + \beta_1 CAT + \beta_2 AGE + \beta_3 ECG$.

EXAMPLE (continued)

(1) CAT = 1, AGE = 40, ECG = 0

$$\text{logit } P(\mathbf{X}) = \alpha + \beta_1 1 + \beta_2 40 + \beta_3 0$$
$$= \boxed{\alpha + \beta_1 + 40\beta_2}$$

(2) CAT = 0, AGE = 40, ECG = 0

$$\text{logit } P(\mathbf{X}) = \alpha + \beta_1 0 + \beta_2 40 + \beta_3 0$$
$$= \boxed{\alpha + 40\beta_2}$$

$$\text{logit } P_1(\mathbf{X}) - \text{logit } P_0(\mathbf{X})$$
$$= (\alpha + \beta_1 + 40\beta_2) - (\alpha + 40\beta_2)$$
$$= \boxed{\beta_1}$$

NOTATION

$\triangle$ = change

$\beta_1 = \triangle \text{ logit}$

$\quad = \triangle \text{ log odds}$ when $\triangle$ CAT = 1   AGE and ECG fixed

$$\text{logit } P(\mathbf{X}) = \alpha + \Sigma \beta_i X_i$$

$i = L$:

$$\boxed{\beta_L = \triangle \ln (\text{odds})}$$

when $= \triangle X_L = 1$, other X's fixed

The first expression below this model shows that when CAT=1, AGE=40, and ECG=0, this logit reduces to $\alpha + \beta_1 + 40\beta_2$.

The second expression shows that when CAT=0, but AGE and ECG remain fixed at 40 and 0, respectively, the logit reduces to $\alpha + 40\beta_2$.

If we subtract the **logit for CAT=0** from the **logit for CAT=1,** after a little arithmetic, we find that the difference is β_1, the coefficient of the variable CAT.

Thus, letting the symbol $\triangle$ denote change, we see that β_1 represents the change in the logit that would result from a unit change in CAT, when the other variables are fixed.

An equivalent explanation is that β_1 represents the *change in the log odds that would result from a one unit change* in the variable CAT when the other variables are fixed. These two statements are equivalent because, by definition, a *logit* is a *log odds*, so that the difference between two logits is the same as the difference between two log odds.

More generally, using the logit expression, if we focus on any coefficient, say β_L, for $i=L$, we can provide the following interpretation:

β_L represents the change in the log odds that would result from a one unit change in the variable X_L, when all other X's are fixed.

SUMMARY

$$\text{logit } P(\mathbf{X})$$

α = background β_i = change in
     log odds   log odds

In summary, by looking closely at the expression for the logit function, we provide some interpretation for the parameters α and β_i in terms of odds, actually *log odds*.

logit ⟶ ? ⟶ OR

Now, how can we use this information about logits to obtain an **odds ratio,** rather than an odds? After all, we are typically interested in measures of association, like odds ratios, when we carry out epidemiologic research.

VIII. Derivation of OR Formula

$$OR = \frac{odds_1}{odds_0}$$

Any **odds ratio,** by definition, is a ratio of two odds, written here as **odds$_1$** divided by **odds$_0$**, in which the subscripts indicate two individuals or two groups of individuals being compared.

EXAMPLE

(1) CAT = 1, AGE = 40, ECG = 0

(0) CAT = 0, AGE = 40, ECG = 0

Now we give an example of an odds ratio in which we compare two groups, called group 1 and group 0. Using our CHD example involving independent variables CAT, AGE, and ECG, group 1 might denote persons with CAT=1, AGE=40, and ECG=0, whereas group 0 might denote persons with CAT=0, AGE=40, and ECG=0.

$\mathbf{X} = (X_1, X_2, \ldots, X_k)$

More generally, when we describe an odds ratio, the two groups being compared can be defined in terms of the bold $\mathbf{X}$ symbol, which denotes a general collection of X variables, from 1 to k.

(1) $\mathbf{X_1} = (X_{11}, X_{12}, \ldots, X_{1k})$

(0) $\mathbf{X_0} = (X_{01}, X_{02}, \ldots, X_{0k})$

Let $\mathbf{X}_1$ denote the collection of X's that specify group 1 and let $\mathbf{X}_0$ denote the collection of X's that specify group 0.

EXAMPLE

$\mathbf{X} = (CAT, AGE, ECG)$

(1) $\mathbf{X}_1 = (CAT = 1, AGE = 40, ECG = 0)$
(0) $\mathbf{X}_0 = (CAT = 0, AGE = 40, ECG = 0)$

In our example, then, k, the number of variables, equals 3, and

$\mathbf{X}$ is the collection of variables CAT, AGE, and ECG,
$\mathbf{X}_1$ corresponds to CAT=1, AGE=40, and ECG=0,
    whereas
$\mathbf{X}_0$ corresponds to CAT=0, AGE=40 and ECG=0.

NOTATION

$$OR_{\mathbf{X}_1, \mathbf{X}_0} = \frac{odds \ for \ \mathbf{X}_1}{odds \ for \ \mathbf{X}_0}$$

Notationally, to distinguish the two groups $\mathbf{X}_1$ and $\mathbf{X}_0$ in an **odds ratio,** we can write $OR_{\mathbf{X}_1, \ \mathbf{X}_0}$ equals the **odds** for $\mathbf{X}_1$ *divided by* the *odds* for $\mathbf{X}_0$.

We will now apply the logistic model to this expression to obtain a general odds ratio formula involving the logistic model parameters.

$$P(\mathbf{X}) = \frac{1}{1 + e^{-(\alpha + \Sigma\beta_i X_i)}}$$

$(1)\ \text{odds}: \dfrac{P(\mathbf{X}_1)}{1 - P(\mathbf{X}_1)}$

$(0)\ \text{odds}: \dfrac{P(\mathbf{X}_0)}{1 - P(\mathbf{X}_0)}$

$$\frac{\text{odds for } \mathbf{X}_1}{\text{odds for } \mathbf{X}_0} = \frac{\dfrac{P(\mathbf{X}_1)}{1 - P(\mathbf{X}_1)}}{\dfrac{P(\mathbf{X}_0)}{1 - P(\mathbf{X}_0)}} = \text{ROR}_{\mathbf{X}_1,\,\mathbf{X}_0}$$

$$\text{ROR} = \frac{\dfrac{P(\mathbf{X}_1)}{1 - P(\mathbf{X}_1)}}{\dfrac{P(\mathbf{X}_0)}{1 - P(\mathbf{X}_0)}} \qquad P(\mathbf{X}) = \frac{1}{1 + e^{-(\alpha + \Sigma\beta_i X_i)}}$$

$(1)\ \dfrac{P(\mathbf{X}_1)}{1 - P(\mathbf{X}_1)} = e^{(\alpha + \Sigma\beta_i X_{1i})}$

$(0)\ \dfrac{P(\mathbf{X}_0)}{1 - P(\mathbf{X}_0)} = e^{(\alpha + \Sigma\beta_i X_{0i})}$

$$\text{ROR}_{\mathbf{X}_1,\mathbf{X}_0} = \frac{\text{odds for } \mathbf{X}_1}{\text{odds for } \mathbf{X}_0} = \frac{e^{(\alpha + \Sigma\beta_i X_{1i})}}{e^{(\alpha + \Sigma\beta_i X_{0i})}}$$

Algebraic theory: $\dfrac{e^a}{e^b} = e^{a-b}$

$a = \alpha + \beta_i X_{1i}, \quad b = \alpha + \beta_i X_{0i}$

Given a logistic model of the general form P(**X**),

we can write the **odds** for **group 1** as $P(\mathbf{X}_1)$ divided by $1 - P(\mathbf{X}_1)$

and the **odds** for **group 0** as $P(\mathbf{X}_0)$ divided by $1 - P(\mathbf{X}_0)$.

To get an odds ratio, we then divide the first odds by the second odds. The result is an expression for the odds ratio written in terms of the two risks $P(\mathbf{X}_1)$ and $P(\mathbf{X}_0)$, that is, $P(\mathbf{X}_1)$ over $1 - P(\mathbf{X}_1)$ divided by $P(\mathbf{X}_0)$ over $1 - P(\mathbf{X}_0)$.

We denote this ratio as **ROR**, for **risk odds ratio**, as the probabilities in the odds ratio are all defined as risks. However, we still do not have a convenient formula.

Now, to obtain a convenient computational formula, we can substitute the mathematical expression 1 over 1 plus e to minus the quantity $(\alpha + \Sigma\beta_i X_i)$ for P(**X**) into the **risk odds ratio** formula above.

For group 1, the **odds** $P(\mathbf{X}_1)$ over $1 - P(\mathbf{X}_1)$ reduces algebraically to e to the linear sum α plus the sum of β_i times X_{1i}, where X_{1i} denotes the value of the variable X_i for group 1.

Similarly, the odds for group 0 reduces to e to the linear sum α plus the sum of β_i times X_{0i}, where X_{0i} denotes the value of variable X_i for group 0.

To obtain the **ROR**, we now substitute in the numerator and denominator the exponential quantities just derived to obtain e to the group 1 linear sum divided by e to the group 0 linear sum.

The above expression is of the form e to the a divided by e to the b, where a and b are linear sums for groups 1 and 0, respectively. From algebraic theory, it then follows that this ratio of two exponentials is equivalent to e to the difference in exponents, or e to the a minus b.

$$\text{ROR} = e^{(\alpha + \Sigma\beta_i X_{1i}) - (\alpha + \Sigma\beta_i X_{0i})}$$

$$= e^{\left[\alpha - \alpha + \Sigma\beta_i (X_{1i} - X_{0i})\right]}$$

$$= e^{\Sigma\beta_i (X_{1i} - X_{0i})}$$

• $$\boxed{\text{ROR}_{\mathbf{X}_1, \mathbf{X}_0} = e^{\sum\limits_{i=1}^{k} \beta_i (X_{1i} - X_{0i})}}$$

$$\boxed{e^{a+b} = e^a \times e^b}$$

$$e^{\sum\limits_{i=1}^{k} z_i} = e^{z_1} \times e^{z_2} \times \cdots e^{z_k}$$

NOTATION

$$= \prod_{i=1}^{k} e^{z_i}$$

$$z_i = \beta_i (X_{1i} - X_{0i})$$

• $$\boxed{\text{ROR}_{\mathbf{X}_1, \mathbf{X}_0} = \prod_{i=1}^{k} e^{\beta_i (X_{1i} - X_{0i})}}$$

$$\prod_{i=1}^{k} e^{\beta_i (X_{1i} - X_{0i})}$$

$$= e^{\beta_1 (X_{11} - X_{01})} e^{\beta_2 (X_{12} - X_{02})} \cdots e^{\beta_k (X_{1k} - X_{0k})}$$

We then find that the **ROR** equals e to the difference between the two linear sums.

In computing this difference, the α's cancel out and the β_i's can be factored for the ith variable.

Thus, the expression for **ROR** simplifies to the quantity e to the sum β_i times the difference between X_{1i} and X_{0i}.

We thus have a general exponential formula for the risk odds ratio from a logistic model comparing any two groups of individuals, as specified in terms of $\mathbf{X}_1$ and $\mathbf{X}_0$. Note that the formula involves the β_i's but not α.

We can give an equivalent alternative to our ROR formula by using the algebraic rule that says that the exponential of a sum is the same as the product of the exponentials of each term in the sum. That is, e to the a plus b equals e to the a times e to the b.

More generally, e to the sum of z_i equals the product of e to the z_i over all i, where the z_i's denote any set of values.

We can alternatively write this expression using the product symbol Π, where Π is a mathematical notation which denotes the product of a collection of terms.

Thus, using algebraic theory and letting z_i correspond to the term β_i times $(X_{1i} - X_{0i})$,

we obtain the **alternative formula** for **ROR** as the product from $i=1$ to k of e to the β_i times the difference $(X_{1i} - X_{0i})$

That is, Π of e to the β_i times $(X_{1i} - X_{0i})$ equals e to the β_1 times $(X_{11} - X_{01})$ multiplied by e to the β_2 times $(X_{12} - X_{02})$ multiplied by additional terms, the final term

being e to the β_k times $(X_{1k} - X_{0k})$.

$$ROR_{\mathbf{X}_1, \mathbf{X}_0} = \prod_{i=1}^{k} e^{\beta_i(X_{1i} - X_{0i})}$$

- Multiplicative

The **product formula** for the **ROR**, shown again here, gives us an interpretation about how each variable in a logistic model contributes to the odds ratio.

In particular, we can see that each of the variables X_i contributes jointly to the odds ratio in a **multiplicative** way.

For example, if

e to the β_i times $(X_{1i} - X_{0i})$ is

3 for variable 2 and

4 for variable 5,

then the joint contribution of these two variables to the odds ratio is **3 × 4**, or **12**.

EXAMPLE

$e^{\beta_2(X_{12} - X_{02})} = 3$

$e^{\beta_5(X_{15} - X_{05})} = 4$

$3 \times 4 = 12$

Logistic model ⇒ multiplicative
             OR formula

Thus, the product or Π formula for **ROR** tells us that, when the logistic model is used, the contribution of the variables to the odds ratio is **multiplicative.**

Other models ⇒ other OR formulae

A model different from the logistic model, depending on its form, might imply a different (for example, an additive) contribution of variables to the odds ratio. An investigator not willing to allow a multiplicative relationship may, therefore, wish to consider other models or other OR formulae. Other such choices are beyond the scope of this presentation.

IX. Example of OR Computation

$$ROR_{\mathbf{X}_1, \mathbf{X}_0} = e^{\sum_{i=1}^{k} \beta_i(X_{1i} - X_{0i})}$$

Given the choice of a logistic model, the version of the formula for the **ROR**, shown here as the exponential of a sum, is the most useful for computational purposes.

EXAMPLE

$\mathbf{X} = (CAT, AGE, ECG)$
(1) $CAT = 1, AGE = 40, ECG = 0$
(0) $CAT = 0, AGE = 40, ECG = 0$

$\mathbf{X}_1 = (CAT = 1, AGE = 40, ECG = 0)$

For example, suppose the **X**'s are CAT, AGE, and ECG, as in our earlier examples.

Also suppose, as before, that we wish to obtain an expression for the odds ratio that compares the following two groups: **group 1** with CAT=1, AGE=40, and ECG=0, and **group 0** with CAT=0, AGE=40, and ECG=0.

For this situation, we let $\mathbf{X}_1$ be specified by CAT=1, AGE=40, and ECG=0,

EXAMPLE (continued)

$X_0 = (CAT = 0, AGE = 40, ECG = 0)$

$ROR_{X_1, X_0} = e^{\sum_{i=1}^{k} \beta_i (X_{1i} - X_{0i})}$

$= e^{\beta_1(1-0) + \beta_2(40-40) + \beta_3(0-0)}$

$= e^{\beta_1 + 0 + 0}$

$= e^{\beta_1}$ ⟵ ————— coefficient of CAT in

logit $P(X) = \alpha + \beta_1 CAT + \beta_2 AGE + \beta_3 ECG$

$ROR_{X_1, X_0} = e^{\beta_1}$

(1) $\boxed{CAT = 1,}$ AGE = 40, ECG = 0
(0) $\boxed{CAT = 0,}$ AGE = 40, ECG = 0

$ROR_{X_1, X_0} = e^{\beta_1}$

= an "adjusted" OR

AGE and ECG:

- fixed
- same
- control variables

e^{β_1}: population ROR

$e^{\hat{\beta}_1}$: estimated ROR

and let X_0 be specified by CAT=0, AGE=40, and ECG=0.

Starting with the general formula for the **ROR**, we then substitute the values for the X_1 and X_0 variables in the formula.

We then obtain **ROR** equals e to the β_1 times $(1 - 0)$ plus β_2 times $(40 - 40)$ plus β_3 times $(0 - 0)$.

The last two terms reduce to 0,

so that our final expression for the **odds ratio** is e to the β_1, where β_1 is the coefficient of the variable CAT.

Thus, for our example, even though the model involves the three variables CAT, ECG, and AGE, the odds ratio expression comparing the two groups involves only the parameter involving the variable CAT. Notice that of the three variables in the model, the variable CAT is the only variable whose value is different in groups 1 and 0. In both groups, the value for AGE is 40 and the value for ECG is 0.

The formula e to the β_1 may be interpreted, in the context of this example, as an **adjusted odds ratio.** This is because we have derived this expression from a logistic model containing two other variables, namely, AGE, and ECG, in addition to the variable CAT. Furthermore, we have fixed the values of these other two variables to be the same for each group. Thus, e to β_1 gives an odds ratio for the effect of the CAT variable **adjusted** for AGE and ECG, where the latter two variables are being treated as **control variables.**

The expression e to the β_1 denotes a population odds ratio parameter because the term β_1 is itself an unknown population parameter.

An estimate of this population odds ratio would be denoted by e to the β_1 "hat." This term, β_1 hat, denotes an **estimate** of β_1 obtained by using some computer package to fit the logistic model to a set of data.

X. Special Case for (0, 1) Variables

Adjusted OR $= e^{\beta}$
where β = coefficient of (0, 1) variable

EXAMPLE

$$\text{logit } P(\mathbf{X}) = \alpha + \beta_1 \boxed{CAT} + \beta_2 AGE + \beta_3 ECG$$

adjusted

X_i(0, 1): adj. ROR $= e^{\beta_i}$

controlling for other X's

EXAMPLE

$$\text{logit } P(\mathbf{X}) = \alpha + \beta_1 CAT + \beta_2 AGE + \beta_3 \boxed{ECG}$$

adjusted

ECG (0, 1): adj. ROR $= e^{\beta_3}$

controlling for CAT and AGE

SUMMARY

X_i is (0, 1): ROR $= e^{\beta_i}$

General OR formula:

$$ROR = e^{\sum_{i=1}^{k} \beta_i \left(X_{1i} - X_{0i} \right)}$$

EXAMPLE

$$\text{logit } P(\mathbf{X}) = \alpha + \beta_1 CAT + \beta_2 AGE + \beta_3 ECG$$

main effect variables

Our example illustrates an important special case of the general odds ratio formula for logistic regression that applies to (0, 1) variables. That is, an **adjusted odds ratio** can be obtained by exponentiating the coefficient of a (0, 1) variable in the model.

In our example, that variable is CAT, and the other two variables, AGE and ECG, are the ones for which we adjusted.

More generally, if the variable of interest is X_i, a (0, 1) variable, then e to the β_i, where β_i is the coefficient of X_i, gives an adjusted odds ratio involving the effect of X_i adjusted or controlling for the remaining X variables in the model.

Suppose, for example, our focus had been on **ECG,** also a (0, 1) variable, instead of on CAT in a logistic model involving the same variables CAT, AGE, and ECG.

Then e to the β_3, where β_3 is the coefficient of ECG, would give the adjusted odds ratio for the effect of ECG, controlling for CAT and AGE.

Thus, we can obtain an adjusted odds ratio for each (0, 1) variable in the logistic model by exponentiating the coefficient corresponding to that variable. This formula is much simpler than the general formula for ROR described earlier.

Note, however, that the example we have considered involves only **main effect variables,** like CAT, AGE and ECG, and that the model does not contain product terms like CAT × AGE or AGE × ECG.

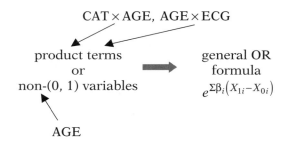

When the model contains product terms, like **CAT** × AGE, or variables that are not (0, 1), like the continuous variable AGE, the simple formula will not work if the focus is on any of these variables. In such instances, we must use the general formula instead.

Chapters

This presentation is now complete. We suggest that you review the material covered here by reading the summary section. You may also want to do the practice exercises and the test which follows. Then continue to the next chapter entitled, "Important Special Cases of the Logistic Model."

Detailed Outline

I. **The multivariable problem** (pages 4–5)
   A. Example of a multivariate problem in epidemiologic research, including the issue of controlling for certain variables in the assessment of an exposure–disease relationship.
   B. The general multivariate problem: assessment of the relationship of several independent variables, denoted as X's, to a dependent variable, denoted as D.
   C. Flexibility in the types of independent variables allowed in most regression situations: A variety of variables is allowed.
   D. Key restriction of model characteristics for the logistic model: The dependent variable is dichotomous.

II. **Why is logistic regression popular?** (pages 5–7)
   A. Description of the logistic function.
   B. Two key properties of the logistic function: Range is between 0 and 1 (good for describing probabilities) and the graph of function is S-shaped (good for describing combined risk factor effect on disease development).

III. **The logistic model** (pages 7–8)
   A. Epidemiologic framework
   B. Model formula: $P(D = 1 \mid X_1, ..., X_k) = P(\mathbf{X})$
   $$= 1/\{1 + \exp[-(\alpha + \Sigma\beta_i X_i)]\}.$$

IV. **Applying the logistic model formula** (pages 9–11)
   A. The situation: independent variables CAT (0, 1), AGE (constant), ECG (0, 1); dependent variable CHD(0, 1); fit logistic model to data on 609 people.
   B. Results for fitted model: estimated model parameters are $\hat{\alpha} = -3.911$, $\hat{\beta}_1$(CAT)=0.65, $\hat{\beta}_2$(AGE)=0.029, and $\hat{\beta}_3$ (ECG)=0.342.
   C. Predicted risk computations:
   $P(\mathbf{X})$ "hat" for CAT=1, AGE=40, ECG=0: 0.1090,
   $P(\mathbf{X})$ "hat" for CAT=0, AGE=40, ECG=0: 0.0600.
   D. Estimated risk ratio calculation and interpretation: 0.1090/0.0600=1.82.
   E. Risk ratio (RR) vs. odds ratio (OR): RR computation requires specifying all X's; OR is more natural measure for logistic model.

V. **Study design issues** (pages 11–15)
   A. Follow-up orientation.
   B. Applicability to case-control and cross-sectional studies? <u>Yes</u>.
   C. Limitation in case-control and cross-sectional studies: cannot estimate risks, but can estimate odds ratios.
   D. The limitation in mathematical terms: for case-control and cross-sectional studies, cannot get a good estimate of the constant.

C. Expressing the risk odds ratio (ROR) in terms of P($\mathbf{X}$):

$$\text{ROR} = \frac{\left(\text{odds for } \mathbf{X}_1\right)}{\left(\text{odds for } \mathbf{X}_0\right)}$$

$$= \frac{P(\mathbf{X}_1)/1 - P(\mathbf{X}_1)}{P(\mathbf{X}_0)/1 - P(\mathbf{X}_0)}.$$

D. Substitution of the model form for P($\mathbf{X}$) in the above ROR formula to obtain general ROR formula:
$$\text{ROR} = \exp[\Sigma\beta_i(X_{1i} - X_{0i})] = \Pi\{\exp[\beta_i(X_{1i} - X_{0i})]\}$$

E. Interpretation from the product (Π) formula: The contribution of each X_i variable to the odds ratio is **multiplicative.**

IX. Example of OR computation (pages 25–26)

A. Example of ROR formula for CAT, AGE, and ECG example using X_1 and X_0 specified in VIII B above:
$$\text{ROR} = \exp(\beta_1), \text{ where } \beta_1 \text{ is the coefficient of CAT.}$$

B. Interpretation of $\exp(\beta_1)$: an adjusted ROR for effect of CAT, controlling for AGE and ECG.

X. Special case for (0, 1) variables (pages 27–28)

A. General rule for (0, 1) variables: If variable is X_i, then ROR for effect of X_i controlling for other X's in model is given by the formula ROR = $\exp(\beta_i)$, where β_i is the coefficient of X_i.

B. Example of formula in A for ECG, controlling for CAT and AGE.

C. Limitation of formula in A: Model can contain only main effect variables for X's, and variable of focus must be (0, 1).

KEY FORMULAE

[$\exp(a) = e^a$ for any number a]

LOGISTIC FUNCTION: $f(z) = 1/[1 + \exp(-z)]$

LOGISTIC MODEL: $P(\mathbf{X}) = 1/\{1 + \exp[-(\alpha + \Sigma\beta_i X_i)]\}$

LOGIT TRANSFORMATION: logit $P(\mathbf{X}) = \alpha + \Sigma\beta_i X_i$

RISK ODDS RATIO (general formula):
$\text{ROR}_{\mathbf{X}_1, \mathbf{X}_0} := \exp[\Sigma\beta_i(X_{1i} - X_{0i})] = \Pi\{\exp[\beta_i(X_{1i} - X_{0i})]\}$

RISK ODDS RATIO [(0, 1) variables]: $\text{ROR} = \exp(\beta_i)$ for the effect of the variable X_i *adjusted* for the other X's

Practice Exercises

Suppose you are interested in describing whether social status, as measured by a (0, 1) variable called SOC, is associated with cardiovascular disease mortality, as defined by a (0, 1) variable called CVD. Suppose further that you have carried out a 12-year follow-up study of 200 men who are 60 years old or older. In assessing the relationship between SOC and CVD, you decide that you want to control for smoking status [SMK, a (0, 1) variable] and systolic blood pressure (SBP, a continuous variable).

In analyzing your data, you decide to fit two logistic models, each involving the dependent variable CVD, but with different sets of independent variables. The variables involved in each model and their estimated coefficients are listed below:

Model 1		Model 2	
VARIABLE	COEFFICIENT	VARIABLE	COEFFICIENT
CONSTANT	−1.1800	CONSTANT	−1.1900
SOC	−0.5200	SOC	−0.5000
SBP	0.0400	SBP	0.0100
SMK	−0.5600	SMK	−0.4200
SOC × SBP	−0.0330		
SOC × SMK	0.1750		

1. For each of the models fitted above, state the form of the logistic model that was used (i.e., state the model in terms of the unknown population parameters and the independent variables being considered).

   Model 1:

   Model 2:

2. For each of the above models, state the form of the estimated model in logit terms.

   Model 1: logit $P(\mathbf{X})=$

   Model 2: logit $P(\mathbf{X})=$

3. Using model 1, compute the estimated risk for CVD death (i.e., CVD=1) for a high social class (SOC=1) smoker (SMK=1) with SBP=150. (You will need a calculator to answer this. If you don't have one, just state the computational formula that is required, with appropriate variable values plugged in.)

4. Using model 2, compute the estimated risk for CVD death for the following two persons:

   Person 1: SOC=1, SMK=1, SBP=150.
   Person 2: SOC=0, SMK=1, SBP=150.
   (As with the previous question, if you don't have a calculator, you may just state the computations that are required.)

   Person 1:

   Person 2:

5. Compare the estimated risk obtained in Exercise 3 with that for person 1 in Exercise 4. Why aren't the two risks exactly the same?

6. Using model 2 results, compute the risk ratio that compares person 1 with person 2. Interpret your answer.

7. If the study design had been either case-control or cross-sectional, could you have legitimately computed risk estimates as you did in the previous exercises? Explain.

8. If the study design had been case-control, what kind of measure of association could you have legitimately computed from the above models?

9. For model 2, compute and interpret the estimated odds ratio for the effect of SOC, controlling for SMK and SBP? (Again, if you do not have a calculator, just state the computations that are required.)

10. Which of the following general formulae is *not* appropriate for computing the effect of SOC controlling for SMK and SBP in *model 1*? (Circle one choice.) Explain your answer.

    a. $\exp(\beta_S)$, where β_S is the coefficient of SOC in model 1.
    b. $\exp[\Sigma\beta_i(X_{1i} - X_{0i})]$.
    c. $\Pi\{\exp[\beta_i(X_{1i} - X_{0i})]\}$.

Test

True or False (Circle T or F)

T   F   1. We can use the logistic model provided all the independent variables in the model are continuous.

T   F   2. Suppose the dependent variable for a certain multivariable analysis is systolic blood pressure, treated continuously. Then, a logistic model should be used to carry out the analysis.

T   F   3. One reason for the popularity of the logistic model is that the range of the logistic function, from which the model is derived, lies between 0 and 1.

T   F   4. Another reason for the popularity of the logistic model is that the shape of the logistic function is linear.

T   F   5. The logistic model describes the probability of disease development, i.e., risk for the disease, for a given a set of independent variables.

T   F   6. The study design framework within which the logistic model is defined is a follow-up study.

T   F   7. Given a fitted logistic model from case-control data, we can estimate the disease risk for a specific individual.

T   F   8. In follow-up studies, we can use a fitted logistic model to estimate a risk ratio comparing two groups provided all the independent variables in the model are specified for both groups.

T   F   9. Given a fitted logistic model from a follow-up study, it is not possible to estimate individual risk as the constant term cannot be estimated.

T   F   10. Given a fitted logistic model from a case-control study, an odds ratio can be estimated.

T   F   11. Given a fitted logistic model from a case-control study, we can estimate a risk ratio if the rare disease assumption is appropriate.

T   F   12. The logit transformation for the logistic model gives the log odds ratio for the comparison of two groups.

T   F   13. The constant term, α, in the logistic model can be interpreted as a baseline log odds for getting the disease.

T   F   14. The coefficient β_i in the logistic model can be interpreted as the change in log odds corresponding to a one unit change in the variable X_i that ignores the contribution of other variables.

T   F   15. We can compute an odds ratio for a fitted logistic model by identifying two groups to be compared in terms of the independent variables in the fitted model.

T   F   16. The product formula for the odds ratio tells us that the joint contribution of different independent variables to the odds ratio is additive.

T   F   17. Given a (0, 1) independent variable and a model containing only main effect terms, the odds ratio that describes the effect of that variable controlling for the others in the model is given by e to the α, where α is the constant parameter in the model.

T   F   18. Given independent variables AGE, SMK [smoking status (0, 1)], and RACE (0, 1), in a logistic model, an adjusted odds ratio for the effect of SMK is given by the natural log of the coefficient for the SMK variable.

T   F   19. Given independent variables AGE, SMK, and RACE, as before, plus the product terms SMK $\times$ RACE and SMK $\times$ AGE, an adjusted odds ratio for the effect of SMK is obtained by exponentiating the coefficient of the SMK variable.

T   F   20. Given the independent variables AGE, SMK, and RACE as in Question 18, but with SMK coded as (1, −1) instead of (0, 1), then e to the coefficient of the SMK variable gives the adjusted odds ratio for the effect of SMK.

21. Which of the following is *not* a property of the logistic model? (Circle one choice.)

   a. The model form can be written as $P(\mathbf{X})=1/\{1+\exp[-(\alpha+\Sigma\beta_i X_i)]\}$, where "exp{·}" denotes the quantity e raised to the power of the expression inside the brackets.

   b. logit $P(\mathbf{X})=\alpha+\Sigma\beta_i X_i$ is an alternative way to state the model.

   c. ROR$=\exp[\Sigma\beta_i(X_{1i}-X_{0i})]$ is a general expression for the odds ratio that compares two groups of $\mathbf{X}$ variables.

   d. ROR$=\Pi\{\exp[\beta_i(X_{1i}-X_{0i})]\}$ is a general expression for the odds ratio that compares two groups of $\mathbf{X}$ variables.

   e. For any variable X_i, ROR$=\exp[\beta_i]$, where β_i is the coefficient of X_i, gives an adjusted odds ratio for the effect of X_i.

Suppose a logistic model involving the variables D=HPT[hypertension status (0, 1)], X_1=AGE(continuous), X_2=SMK(0, 1), X_3=SEX(0, 1), X_4=CHOL (cholesterol level, continuous), and X_5=OCC[occupation (0, 1)] is fit to a set of data. Suppose further that the estimated coefficients of each of the variables in the model are given by the following table:

VARIABLE	COEFFICIENT
CONSTANT	−4.3200
AGE	0.0274
SMK	0.5859
SEX	1.1523
CHOL	0.0087
OCC	−0.5309

22. State the form of the logistic model that was fit to these data (i.e., state the model in terms of the unknown population parameters and the independent variables being considered).

23. State the form of the *estimated* logistic model obtained from fitting the model to the data set.

24. State the estimated logistic model in logit form.

25. Assuming the study design used was a follow-up design, compute the estimated risk for a 40-year-old male (SEX=1) smoker (SMK=1) with CHOL=200 and OCC=1. (You need a calculator to answer this question.)

26. Again assuming a follow-up study, compute the estimated risk for a 40-year-old male nonsmoker with CHOL=200 and OCC=1. (You need a calculator to answer this question.)

27. Compute and interpret the estimated risk ratio that compares the risk of a 40-year-old male smoker to a 40-year-old male nonsmoker, both of whom have CHOL=200 and OCC=1.

28. Would the risk ratio computation of Question 27 have been appropriate if the study design had been either cross-sectional or case-control? Explain.

29. Compute and interpret the estimated odds ratio for the effect of SMK controlling for AGE, SEX, CHOL, and OCC. (If you do not have a calculator, just state the computational formula required.)

30. What assumption will allow you to conclude that the estimate obtained in Question 29 is approximately a risk ratio estimate?

31. If you could not conclude that the odds ratio computed in Question 29 is approximately a risk ratio, what measure of association is appropriate? Explain briefly.

32. Compute and interpret the estimated odds ratio for the effect of OCC controlling for AGE, SMK, SEX, and CHOL. (If you do not have a calculator, just state the computational formula required.)

33. State two characteristics of the variables being considered in this example that allow you to use the $\exp(\beta_i)$ formula for estimating the effect of OCC controlling for AGE, SMK, SEX, and CHOL.

34. Why can you not use the formula $\exp(\beta_i)$ formula to obtain an adjusted odds ratio for the effect of AGE, controlling for the other four variables?

Answers to Practice Exercises

1. *Model 1:* P(**X**)=1/{1+exp−[−1.18−0.52(SOC)+0.04(SBP)−0.56(SMK)
    −0.033(SOC×SBP)+0.175(SOC×SMK)]}.

    Model 2: P(**X**)=1/(1+exp{−[−1.19−0.50(SOC)+0.01(SBP)+0.42(SMK)]}).

2. *Model 1:* logit P(**X**) =−1.18−0.52(SOC)+0.04(SBP)−0.56(SMK)
    −0.033(SOC×SBP)+0.175(SOC×SMK).

    Model 2: logit P(**X**)=−1.19−0.50(SOC)+0.01(SBP)−0.42(SMK).

3. For SOC=1, SBP=150, and SMK=1,
    X=(SOC, SBP, SMK, SOC×SBP, SOC×SMK)=(1, 150, 1, 150, 1) and

    model 1 P(**X**)=1/(1+exp{−[−1.18−0.52(1)+0.04(150)−0.56(1)
    −0.033(1×150)−0.175(1×1)]}).
    =1/{1+exp[−(−1.035)]}
    =1/(1+2.815)
    =0.262

4. For *model 2, person 1* (SOC=1, SMK=1, SBP=150):

    P(**X**)=1/(1+exp{−[−1.19 − 0.50(1) + 0.01 (150) − 0.47(1)]})
    =1/{1 + exp [− (−0.61)]}
    =1/(1 + 1.84)
    =0.352

    For *model 2, person 2* (SOC=0, SMK=1, SBP=150):

    P(**X**)=1/(1 + exp{−[−1.19 − 0.50(0) − 0.01(150) + 0.42(1)]})
    =1/{1 + exp[−(−0.11)]}
    =1/(1 + 1.116)
    =0.473

5. The risk computed for *model 1* is 0.262 whereas the risk computed for *model 2, person 1* is 0.352. Note that both risks are computed for the same person (i.e., SOC=1, SMK=150, SBP=150), yet they yield different values because the models are different. In particular, *model 1* contains two product terms that are not contained in *model 2*, and consequently, computed risks for a given person can be expected to be somewhat different for different models.

6. Using *model 2* results,

$$RR(1 \text{ vs. } 2) = \frac{P(SOC=0, SMK=1, SBP=150)}{P(SOC=1, SMK=1, SBP=150)}$$

$$= 0.352/0.473 = 1/1.34 = 0.744$$

This estimated risk ratio is less than 1 because the risk for high social class persons (SOC=1) is less than the risk for low social class persons (SOC=0) in this data set. More specifically, the risk for low social class persons is 1.34 times as large as the risk for high social class persons.

7. No. If the study design had been either case control or cross sectional, risk estimates could not be computed because the constant term (α) in the model could not be estimated. In other words, even if the computer printed out values of -1.18 or -1.19 for the constant terms, these numbers would not be legitimate estimates of α.

8. For case-control studies, only odds ratios, not risks or risk ratios, can be computed directly from the fitted model.

9. $\widehat{OR}$(SOC=1 vs. SOC=0 controlling for SMK and SBP)

$= e^{\hat{\beta}}$, where $\hat{\beta} = -0.50$ is the estimated coefficient of SOC in the fitted model

$= \exp(-0.50)$
$= 0.6065 = 1/1.65.$

The estimated odds ratio is less than 1, indicating that, for this data set, the risk of CVD death for high social class persons is less than the risk for low social class persons. In particular, the risk for low social class persons is estimated as 1.65 times as large as the risk for high social class persons.

10. Choice (a) is *not* appropriate for the effect of SOC using model 1. Model 1 contains interaction terms, whereas choice (a) is appropriate only if all the variables in the model are main effect terms. Choices (b) and (c) are two equivalent ways of stating the general formula for calculating the odds ratio for any kind of logistic model, regardless of the types of variables in the model.

2

Important Special Cases of the Logistic Model

Introduction

In this chapter, several important special cases of the logistic model involving a single (0, 1) exposure variable are considered with their corresponding odds ratio expressions. In particular, focus is on defining the independent variables that go into the model and on computing the odds ratio for each special case. Models that account for the potential confounding effects and potential interaction effects of covariates are emphasized.

Abbreviated Outline

The outline below gives the user a preview of the material to be covered by the presentation. A detailed outline for review purposes follows the presentation.

    I. **Overview (page 42)**
    II. **Special case—Simple analysis (pages 43–46)**
    III. **Assessing multiplicative interaction (pages 46–52)**
    IV. **The *E, V, W* model—A general model containing a (0, 1) exposure and potential confounders and effect modifiers (pages 52–61)**
    V. **Logistic model for matched data (pages 61–64)**

Objectives

Upon completion of this chapter, the learner should be able to:

1.   State or recognize the logistic model for a simple analysis.
2.   Given a model for simple analysis:
    a. state an expression for the odds ratio describing the exposure–disease relationship;
    b. state or recognize the null hypothesis of no exposure–disease relationship in terms of parameter(s) of the model;
    c. compute or recognize an expression for the risk for exposed or unexposed persons separately;
    d. compute or recognize an expression for the odds of getting the disease for exposed or unexposed persons separately.
3.   Given two (0, 1) independent variables:
    a. state or recognize a logistic model which allows for the assessment of interaction on a multiplicative scale;
    b. state or recognize the expression for no interaction on a multiplicative scale in terms of odds ratios for different combinations of the levels of two (0, 1) independent variables;
    c. state or recognize the null hypothesis for no interaction on a multiplicative scale in terms of one or more parameters in an appropriate logistic model.

4. Given a study situation involving a (0, 1) exposure variable and several control variables:

    a. state or recognize a logistic model which allows for the assessment of the exposure–disease relationship, controlling for the potential confounding and potential interaction effects of functions of the control variables;

    b. compute or recognize the expression for the odds ratio for the effect of exposure on disease status adjusting for the potential confounding and interaction effects of the control variables in the model;

    c. state or recognize an expression for the null hypothesis of no interaction effect involving one or more of the effect modifiers in the model;

    d. assuming no interaction, state or recognize an expression for the odds ratio for the effect of exposure on disease status adjusted for confounders;

    e. assuming no interaction, state or recognize the null hypothesis for testing the significance of this odds ratio in terms of a parameter in the model.

5. Given a logistic model involving interaction terms, state or recognize that the expression for the odds ratio will give different values for the odds ratio depending on the values specified for the effect modifiers in the model.

6. Given a study situation involving matched case-control data:

    a. state or recognize a logistic model for the analysis of matched data that controls for both matched and unmatched variables;

    b. state how matching is incorporated into a logistic model using dummy variables;

    c. state or recognize the expression for the odds ratio for the exposure–disease effect that controls for both matched and unmatched variables;

    d. state or recognize null hypotheses for no interaction effect, or for no exposure–disease effect, given no interaction effect.

Presentation

I. Overview

Special Cases:

$$\begin{pmatrix} \begin{array}{|c|c|} \hline a & b \\ \hline c & d \\ \hline \end{array} \end{pmatrix}$$

- Simple analysis

- Multiplicative interaction

- Controlling several confounders and effect modifiers

- Matched data

General logistic model formula:

$$P(\mathbf{X}) = \frac{1}{1 + e^{-(\alpha + \Sigma \beta_i X_i)}}$$

$$\mathbf{X} = (X_1, X_2, \ldots, X_k)$$

α, β_i = unknown parameters

D = dichotomous outcome

$$\text{logit } P(\mathbf{X}) = \underbrace{\alpha + \sum \beta_i X_i}_{\text{linear sum}}$$

$$\text{ROR} = e^{\sum\limits_{i=1}^{k} \beta_i (X_{1i} - X_{0i})}$$

$$= \prod_{i=1}^{k} e^{\beta_i (X_{1i} - X_{0i})}$$

This presentation describes important special cases of the general logistic model when there is a single (0, 1) exposure variable. Special case models include simple analysis of a fourfold table; assessment of multiplicative interaction between two dichotomous variables; controlling for several confounders and interaction terms; and analysis of matched data. In each case, we consider the definitions of variables in the model and the formula for the odds ratio describing the exposure–disease relationship.

Recall that the general logistic model for k independent variables may be written as $P(\mathbf{X})$ equals 1 over 1 plus e to minus the quantity α plus the sum of $\beta_i X_i$, where $P(\mathbf{X})$ denotes the probability of developing a disease of interest given values of a collection of independent variables X_1, X_2, through X_k, that are collectively denoted by the **bold X.** The terms α and β_i in the model represent unknown parameters which we need to estimate from data obtained for a group of subjects on the X's and on D, a dichotomous disease outcome variable.

An alternative way of writing the logistic model is called the logit form of the model. The expression for the logit form is given here.

The general odds ratio formula for the logistic model is given by either of two formulae. The first formula is of the form e to a sum of linear terms. The second is of the form of the product of several exponentials; that is, each term in the product is of the form e to some power. Either formula requires two specifications $\mathbf{X}_1$ and $\mathbf{X}_0$ of the collection of k independent variables X_1, X_2, . . ., X_k.

We now consider a number of important special cases of the logistic model and their corresponding odds ratio formulae.

II. Special Case—Simple Analysis

$X_1 = E$ = exposure (0, 1)

D = disease (0, 1)

We begin with the simple situation involving one dichotomous independent variable, which we will refer to as an **exposure** variable and will denote it as $X_1 = E$. Because the disease variable, D, considered by a logistic model is dichotomous, we can use a two-way table with four cells to characterize this analysis situation, which is often referred to as a **simple analysis.**

	$E=1$	$E=0$
$D=0$	a	b
$D=1$	c	d

(whatever)

For convenience, we define the exposure variable as a (0, 1) variable and place its values in the two columns of the table. We also define the disease variable as a (0, 1) variable and place its values in the rows of the table. The cell frequencies within the fourfold table are denoted as *a*, *b*, *c*, and *d*, as is typically presented for such a table.

$$P(\mathbf{X}) = \frac{1}{1 + e^{-(\alpha + \beta_1 E)}}$$

where $E = (0, 1)$ variable

Note: Other coding schemes

$(1, -1), (1, 2), (2, 1)$

A logistic model for this simple analysis situation can be defined by the expression $P(\mathbf{X})$ equals 1 over 1 plus *e* to minus the quantity α plus β_1 times *E*, where *E* takes on the value 1 for exposed persons and 0 for unexposed persons. Note that other coding schemes for *E* are also possible, such as $(1, -1)$, $(1, 2)$, or even $(2, 1)$. However, we defer discussing such alternatives until Chapter 3.

$$\text{logit } P(\mathbf{X}) = \alpha + \beta_1 E$$

The logit form of the logistic model we have just defined is of the form logit $P(\mathbf{X})$ equals the simple linear sum α plus β_1 times *E*. As stated earlier in our review, this logit form is an alternative way to write the statement of the model we are using.

$$P(\mathbf{X}) = \Pr(D = 1 | E)$$

$$E = 1: R_1 = \Pr(D = 1 | E = 1)$$

$$E = 0: R_0 = \Pr(D = 1 | E = 0)$$

The term $P(\mathbf{X})$ for the simple analysis model denotes the probability that the disease variable D takes on the value 1, given whatever the value is for the exposure variable E. In epidemiologic terms, this probability denotes the **risk** for developing the disease, given exposure status. When the value of the exposure variable equals 1, we call this risk $\mathbf{R}_1$, which is the conditional probability that D equals 1 given that E equals 1. When E equals 0, we denote the risk by $\mathbf{R}_0$, which is the conditional probability that D equals 1 given that E equals 0.

$$\text{ROR}_{E=1 \text{ vs. } E=0} = \frac{\dfrac{R_1}{1-R_1}}{\dfrac{R_0}{1-R_0}}$$

We would like to use the above model for simple analysis to obtain an expression for the odds ratio that compares exposed persons with unexposed persons. Using the terms R_1 and R_0, we can write this odds ratio as R_1 divided by 1 minus R_1 over R_0 divided by 1 minus R_0.

Substitute $P(X) = \dfrac{1}{1+e^{-(\alpha+\Sigma\beta_i X_i)}}$

into ROR formula:

$$E=1: \quad R_1 = \frac{1}{1+e^{-(\alpha+[\beta_1 \times 1])}}$$

$$= \frac{1}{1+e^{-(\alpha+\beta_1)}}$$

To compute the odds ratio in terms of the parameters of the logistic model, we substitute the logistic model expression into the odds ratio formula.

For E equal to 1, we can write R_1 by substituting the value E equals 1 into the model formula for $P(X)$. We then obtain 1 over 1 plus e to minus the quantity α plus β_1 times 1, or simply 1 over 1 plus e to minus α plus β_1.

$$E=0: \quad R_0 = \frac{1}{1+e^{-(\alpha+[\beta_1 \times 0])}}$$

$$= \frac{1}{1+e^{-\alpha}}$$

For E equal to zero, we write R_0 by substituting E equal to 0 into the model formula, and we obtain 1 over 1 plus e to minus α.

$$\text{ROR} = \frac{\dfrac{R_1}{1-R_1}}{\dfrac{R_0}{1-R_0}} = \frac{\dfrac{1}{1+e^{-(\alpha+\beta_1)}}}{\dfrac{1}{1+e^{-\alpha}}}$$

algebra

$$= \boxed{e^{\beta_1}}$$

To obtain ROR then, we replace R_1 with 1 over 1 plus e to minus α plus β_1, and we replace R_0 with 1 over 1 plus e to minus α. The ROR formula then simplifies algebraically to e to the β_1, where β_1 is the coefficient of the exposure variable.

General ROR formula used for other special cases

We could have obtained this expression for the odds ratio using the general formula for the ROR that we gave during our review. We will use the general formula now. Also, for other special cases of the logistic model, we will use the general formula rather than derive an odds ratio expression separately for each case.

General:

$$\text{ROR}_{\mathbf{X}_1, \mathbf{X}_0} = e^{\sum\limits_{i=1}^{k} \beta_i \left(X_{1i} - X_{0i} \right)}$$

Simple analysis:

$$k = 1, \quad \mathbf{X} = \left(X_1 \right), \quad \beta_i = \beta_1$$

group 1: $\mathbf{X}_1 = E = 1$
group 0: $\mathbf{X}_0 = E = 0$

$$\mathbf{X}_1 = \left(X_{11} \right) = \left(1 \right)$$
$$\mathbf{X}_0 = \left(X_{01} \right) = \left(0 \right)$$

The general formula computes ROR as e to the sum of each β_i times the difference between X_{1i} and X_{0i}, where X_{1i} denotes the value of the ith X variable for group 1 persons and X_{0i} denotes the value of the ith X variable for group 0 persons. In a simple analysis, we have only one X and one β; in other words, k, the number of variables in the model, equals 1.

For a simple analysis model, group 1 corresponds to exposed persons, for whom the variable X_1, in this case E, equals 1. Group 0 corresponds to unexposed persons, for whom the variable X_1 or E equals 0. Stated another way, for group 1, the collection of X's denoted by the **bold X** can be written as **bold X_1** and equals the collection of one value X_{11}, which equals 1. For group 0, the collection of X's denoted by the **bold X** is written as **bold X_0** and equals the collection of one value X_{01}, which equals 0.

$$\begin{aligned}
\text{ROR}_{\mathbf{X}_1, \mathbf{X}_0} &= e^{\beta_1 \left(X_{11} - X_{01} \right)} \\
&= e^{\beta_1 \left(1 - 0 \right)} \\
&= e^{\beta_1}
\end{aligned}$$

Substituting the particular values of the one X variable into the general odds ratio formula then gives e to the β_1 times the quantity X_{11} minus X_{01}, which becomes e to the β_1 times 1 minus 0, which reduces to e to the β_1.

SIMPLE ANALYSIS SUMMARY

$$P\left(\mathbf{X} \right) = \frac{1}{1 + e^{-\left(\alpha + \beta_1 E \right)}}$$

$$\text{ROR} = e^{\beta_1}$$

In summary, for the simple analysis model involving a (0, 1) exposure variable, the logistic model $P(\mathbf{X})$ equals 1 over 1 plus e to minus the quantity α plus β_1 times E, and the odds ratio which describes the effect of the exposure variable is given by e to the β_1, where β_1 is the coefficient of the exposure variable.

$$\widehat{\text{ROR}}_{\mathbf{X}_1, \mathbf{X}_0} = e^{\hat{\beta}_1}$$

We can estimate this odds ratio by fitting the simple analysis model to a set of data. The estimate of the parameter β_1 is typically denoted as β_1 "hat." The odds ratio estimate then becomes e to the β_1 "hat."

$$E=1 \quad E=0$$

	E=1	E=0
D=1	a	b
D=0	c	d

$$\widehat{ROR} = e^{\hat{\beta}} = ad/bc$$

Simple analysis: does not need computer

Other special cases: require computer

The reader should not be surprised to find out that an alternative formula for the estimated odds ratio for the simple analysis model is the familiar a times d over b times c, where a, b, c, and d are the cell frequencies in the fourfold table for simple analysis. That is, e to the β_1 "hat" obtained from fitting a logistic model for simple analysis can alternatively be computed as ad divided by bc from the cell frequencies of the fourfold table.

Thus, in the simple analysis case, we need not go to the trouble of fitting a logistic model to get an odds ratio estimate as the typical formula can be computed without a computer program. We have presented the logistic model version of simple analysis to show that the logistic model incorporates simple analysis as a special case. More complicated special cases, involving more than one independent variable, require a computer program to compute the odds ratio.

III. Assessing Multiplicative Interaction

We will now consider how the logistic model allows the assessment of interaction between two independent variables.

$$X_1 = A = (0, 1) \text{ variable}$$

$$X_2 = B = (0, 1) \text{ variable}$$

Interaction: equation involving RORs for combinations of A and B

Consider, for example, two $(0, 1)$ X variables, X_1 and X_2, which for convenience we rename as A and B, respectively. We first describe what we mean conceptually by interaction between these two variables. This involves an equation involving risk odds ratios corresponding to different combinations of A and B. The odds ratios are defined in terms of risks, which we now describe.

$$R_{AB} = \text{risk given } A, B$$
$$= \Pr(D = 1 \mid A, B)$$

Let R_{AB} denote the risk for developing the disease, given specified values for A and B; in other words, R_{AB} equals the conditional probability that D equals 1, given A and B.

	B=1	B=0
A=1	R_{11}	R_{10}
A=0	R_{01}	R_{00}

Note: above table not for simple analysis.

Because A and B are dichotomous, there are four possible values for R_{AB}, which are shown in the cells of a two-way table. When A equals 1 and B equals 1, the risk R_{AB} becomes R_{11}. Similarly, when A equals 1 and B equals 0, the risk becomes R_{10}. When A equals 0 and B equals 1, the risk is R_{01}, and finally, when A equals 0 and B equals 0, the risk is R_{00}.

	$B=1$	$B=0$
$A=1$	R_{11}	R_{10}
$A=0$	R_{01}	R_{00}

obvious!

Note that the two-way table presented here does not describe a simple analysis because the row and column headings of the table denote two independent variables rather than one independent variable and one disease variable. Moreover, the information provided within the table is a collection of four risks corresponding to different combinations of both independent variables, rather than four cell frequencies corresponding to different exposure–disease combinations.

	$B=1$	$B=0$
$A=1$		
$A=0$		← referent cell

OR_{ij} = odds ratio for $A=i$ and $B=j$

for the dichotomous dependent variable = 1

OR_{11} = odds $(1, 1)/$odds$(0, 0)$

OR_{10} = odds $(1, 0)/$odds$(0, 0)$

OR_{01} = odds $(0, 1)/$odds$(0, 0)$

Within this framework, odds ratios can be defined to compare the odds for any one cell in the two-way table of risks with the odds for any other cell. In particular, three odds ratios of typical interest compare each of three of the cells to a **referent cell.** The referent cell is usually selected to be the combination A equals 0 and B equals 0. The three odds ratios are then defined as OR_{11}, OR_{10}, and OR_{01}, where OR_{11} equals the odds for cell 11 divided by the odds for cell 00, OR_{10} equals the odds for cell 10 divided by the odds for cell 00, and OR_{01} equals the odds for cell 01 divided by the odds for cell 00.

$$\text{odds}\,(A,B) = R_{AB}/(1-R_{AB})$$

$$OR_{11} = \frac{R_{11}/(1-R_{11})}{R_{00}/(1-R_{00})} = \frac{R_{11}(1-R_{00})}{R_{00}(1-R_{11})}$$

$$OR_{10} = \frac{R_{10}/(1-R_{10})}{R_{00}/(1-R_{00})} = \frac{R_{10}(1-R_{00})}{R_{00}(1-R_{10})}$$

$$OR_{01} = \frac{R_{01}/(1-R_{01})}{R_{00}/(1-R_{00})} = \frac{R_{01}(1-R_{00})}{R_{00}(1-R_{01})}$$

As the odds for any cell A,B is defined in terms of risks as R_{AB} divided by 1 minus R_{AB}, we can obtain the following expressions for the three odds ratios: OR_{11} equals the product of R_{11} times 1 minus R_{00} divided by the product of R_{00} times 1 minus R_{11}. The corresponding expressions for OR_{10} and OR_{01} are similar, where the subscript 11 in the numerator and denominator of the 11 formula is replaced by 10 and 01, respectively.

$$OR_{AB} = \frac{R_{AB}(1-R_{00})}{R_{00}(1-R_{AB})}$$

$A = 0, 1; \quad B = 0, 1$

In general, without specifying the value of A and B, we can write the odds ratio formulae as OR_{AB} equals the product of R_{AB} and 1 minus R_{00} divided by the product of R_{00} and $1 - R_{AB}$, where A takes on the values 0 and 1 and B takes on the values 0 and 1.

DEFINITION

$$OR_{11} = OR_{10} \times OR_{01}$$

no interaction   ↑
on a   multiplication
multiplicative
scale

Now that we have defined appropriate odds ratios for the two independent variables situation, we are ready to provide an equation for assessing interaction. The equation is stated as OR_{11} equals the product of OR_{10} and OR_{01}. If this expression is satisfied for a given study situation, we say that there is "no interaction on a *multiplicative* scale." In contrast, if this expression is not satisfied, we say that there is evidence of interaction on a multiplicative scale.

Note that the right-hand side of the "no interaction" expression requires **multiplication** of two odds ratios, one corresponding to the combination 10 and the other to the combination 01. Thus, the scale used for assessment of interaction is called multiplicative.

No interaction:

$$\begin{pmatrix} \text{effect of } A \text{ and } B \\ \text{acting together} \end{pmatrix} = \begin{pmatrix} \text{combined effect} \\ \text{of } A \text{ and } B \\ \text{acting separately} \end{pmatrix}$$

          ↑   ↑
     OR_{11} $OR_{10} \times OR_{01}$
                              multiplicative
                              scale

When the no interaction equation is satisfied, we can interpret the effect of both variables A and B acting together as being the same as the combined effect of each variable acting separately.

The effect of both variables acting together is given by the odds ratio OR_{11} obtained when A and B are both present, that is, when A equals 1 and B equals 1.

The effect of A acting separately is given by the odds ratio for A equals 1 and B equals 0, and the effect of B acting separately is given by the odds ratio for A equals 0 and B equals 1. The combined separate effects of A and B are then given by the product OR_{10} times OR_{01}.

no interaction formula:

$$OR_{11} = OR_{10} \times OR_{01}$$

Thus, when there is no interaction on a multiplicative scale, OR_{11} equals the product of OR_{10} and OR_{01}.

EXAMPLE

	$B=1$	$B=0$
$A=1$	$R_{11}=0.0350$	$R_{10}=0.0175$
$A=0$	$R_{01}=0.0050$	$R_{00}=0.0025$

$$OR_{11} = \frac{0.0350(1-0.0025)}{0.0025(1-0.0350)} = 14.4$$

$$OR_{10} = \frac{0.0175(1-0.0025)}{0.0025(1-0.0175)} = 7.2$$

$$OR_{01} = \frac{0.0050(1-0.0025)}{0.0025(1-0.0050)} = 2.0$$

$$OR_{11} \overset{?}{=} OR_{10} \times OR_{01}$$

$$14.4 \overset{?}{=} \underset{\underset{14.4}{\uparrow}}{7.2 \times 2.0}$$

$\boxed{\text{Yes}}$

	$B=1$	$B=0$
$R_{11}=0.0700$	$R_{10}=0.0175$
$R_{01}=0.0050$	$R_{00}=0.0025$

$$OR_{11} = 30.0$$

$$OR_{10} = 7.2$$

$$OR_{01} = 2.0$$

$$OR_{11} \overset{?}{=} OR_{10} \times OR_{01}$$

$$30.0 \overset{?}{=} \underset{\uparrow}{7.2 \times 2.0}$$

$\boxed{\text{No}}$

As an example of no interaction on a multiplicative scale, suppose the risks R_{AB} in the fourfold table are given by R_{11} equal to 0.0350, R_{10} equal to 0.0175, R_{01} equal to 0.0050, and R_{00} equal to 0.0025. Then the corresponding three odds ratios are obtained as follows: OR_{11} equals 0.0350 times 1 minus 0.0025 divided by the product of 0.0025 and 1 minus 0.0350, which becomes 14.4; OR_{10} equals 0.0175 times 1 minus 0.0025 divided by the product of 0.0025 and 1 minus 0.0175, which becomes 7.2; and OR_{01} equals 0.0050 times 1 minus 0.0025 divided by the product of 0.0025 and 1 minus 0.0050, which becomes 2.0.

To see if the no interaction equation is satisfied, we check whether OR_{11} equals the product of OR_{10} and OR_{01}. Here we find that OR_{11} equals 14.4 and the product of OR_{10} and OR_{01} is 7.2 times 2, which is also 14.4. Thus, the no interaction equation is satisfied.

In contrast, using a different example, if the risk for the 11 cell is 0.0700, whereas the other three risks remained at 0.0175, 0.0050, and 0.0025, then the corresponding three odds ratios become OR_{11} equals 30.0, OR_{10} equals 7.2, and OR_{01} equals 2.0. In this case, the no interaction equation is not satisfied because the left-hand side equals 30 and the product of the two odds ratios on the right-hand side equals 14. Here, then, we would conclude that there is interaction because the effect of both variables acting together is twice the combined effect of the variables acting separately.

EXAMPLE (continued)

Note: "=" means approximately equal ($\approx$)

e.g., $14.5 \approx 14.0 \Rightarrow$ no interaction

Note that in determining whether or not the no interaction equation is satisfied, the left- and right-hand sides of the equation do not have to be exactly equal. If the left-hand side is approximately equal to the right-hand side, we can conclude that there is no interaction. For instance, if the left-hand side is 14.5 and the right-hand side is 14, this would typically be close enough to conclude that there is no interaction on a multiplicative scale.

REFERENCE

multiplicative interaction vs. additive interaction
Epidemiologic Research, Chapter 19

A more complete discussion of interaction, including the distinction between **multiplicative interaction** and **additive interaction,** is given in Chapter 19 of *Epidemiologic Research* by Kleinbaum, Kupper, and Morgenstern (Van Nostrand Reinhold, New York, 1982).

Logistic model variables:

$$\left.\begin{aligned} X_1 &= A_{(0,\,1)} \\ X_2 &= B_{(0,\,1)} \end{aligned}\right\}\text{main effects}$$

$X_3 = A \times B$ interaction effect variable

We now define a logistic model which allows the assessment of multiplicative interaction involving two $(0, 1)$ indicator variables A and B. This model contains three independent variables, namely, X_1 equal to A, X_2 equal to B, and X_3 equal to the product term A times B. The variables A and B are called main effect variables and the product term is called an interaction effect variable.

$$\text{logit } P(\mathbf{X}) = \alpha + \beta_1 A + \beta_2 B + \beta_3 A \times B$$

where

$$P(\mathbf{X}) = \text{risk given } A \text{ and } B$$

$$= R_{AB}$$

The logit form of the model is given by the expression logit of $P(\mathbf{X})$ equals α plus β_1 times A plus β_2 times B plus β_3 times A times B. $P(\mathbf{X})$ denotes the risk for developing the disease given values of A and B, so that we can alternatively write $P(\mathbf{X})$ as R_{AB}.

$$\beta_3 = \ln_e\left[\frac{\text{OR}_{11}}{\text{OR}_{10} \times \text{OR}_{01}}\right]$$

For this model, it can be shown mathematically that the coefficient β_3 of the product term can be written in terms of the three odds ratios we have previously defined. The formula is β_3 equals the natural log of the quantity OR_{11} divided by the product of OR_{10} and OR_{01}. We can make use of this formula to test the null hypothesis of no interaction on a multiplicative scale.

H_0 no interaction on a multiplicative scale

$\Leftrightarrow H_0 : OR_{11} = OR_{10} \times OR_{01}$

$\Leftrightarrow H_0 : \dfrac{OR_{11}}{OR_{10} \times OR_{01}} = 1$

$\Leftrightarrow H_0 : \ln_e\left(\dfrac{OR_{11}}{OR_{10} \times OR_{01}}\right) = \ln_e 1$

$\Leftrightarrow H_0 : \beta_3 = 0$

$\text{logit } P(\mathbf{X}) = \alpha + \beta_1 A + \beta_2 B + \beta_3 AB$

$H_0 :$ no interaction $\Leftrightarrow \beta_3 = 0$

Test result		*Model*
not significant	$\Rightarrow$	$\alpha + \beta_1 A + \beta_2 B$
significant	$\Rightarrow$	$\alpha + \beta_1 A + \beta_2 B + \beta_3 AB$

MAIN POINT:
Interaction test $\Rightarrow$ test for product terms

One way to state this null hypothesis, as described earlier in terms of odds ratios, is OR_{11} equals the product of OR_{10} and OR_{01}. Now it follows algebraically that this odds ratio expression is equivalent to saying that the quantity OR_{11} divided by OR_{10} times OR_{01} equals 1, or equivalently, that the natural log of this expression equals the natural log of 1, or, equivalently, that β_3 equals 0. Thus, the null hypothesis of no interaction on a multiplicative scale can be equivalently stated as β_3 equals 0.

In other words, a test for the no interaction hypotheses can be obtained by testing for the significance of the coefficient of the product term in the model. If the test is not significant, we would conclude that there is no interaction on a multiplicative scale and we would reduce the model to a simpler one involving only main effects. In other words, the reduced model would be of the form logit $P(\mathbf{X})$ equals α plus β_1 times A plus β_2 times B. If, on the other hand, the test is significant, the model would retain the β_3 term and we would conclude that there is significant interaction on a multiplicative scale.

A description of methods for testing hypotheses for logistic regression models is beyond the scope of this presentation (see Chapter 5.) The main point here is that we can test for interaction in a logistic model by testing for significance of product terms that reflect interaction effects in the model.

EXAMPLE

Case-control study

$ASB = (0, 1)$ variable for asbestos exposure

$SMK = (0, 1)$ variable for smoking status

$D = (0, 1)$ variable for bladder cancer status

As an example of a test for interaction, we consider a study that looks at the combined relationship of asbestos exposure and smoking to the development of bladder cancer. Suppose we have collected case-control data on several persons with the same occupation. We let **ASB** denote a (0, 1) variable indicating asbestos exposure status, **SMK** denote a (0, 1) variable indicating smoking status, and D denote a (0, 1) variable for bladder cancer status.

EXAMPLE (continued)

$$\text{logit}\left(\mathbf{X}\right) = \alpha + \beta_1 \text{ASB} + \beta_2 \text{SMK}$$
$$+ \beta_3 \text{ASB} \times \text{SMK}$$

H_0 : no interaction (multiplicative)

$\Leftrightarrow H_0 : \beta_3 = 0$

Test Result	Conclusion
Not Significant	No interaction on multiplicative scale
Significant ($\hat{\beta}_3 > 0$)	Joint effect > combined effect
Significant ($\hat{\beta}_3 < 0$)	Joint effect < combined effect

or both how ?
down last thru.

To assess the extent to which there is a multiplicative interaction between asbestos exposure and smoking, we consider a logistic model with ASB and SMK as main effect variables and the product term ASB times SMK as an interaction effect variable. The model is given by the expression logit P(**X**) equals α plus β_1 times ASB plus β_2 times SMK plus β_3 times ASB times SMK. With this model, a test for no interaction on a multiplicative scale is equivalent to testing the null hypothesis that β_3, the coefficient of the product term, equals 0.

If this test is not significant, then we would conclude that the effect of asbestos and smoking acting together is equal, on a multiplicative scale, to the combined effect of asbestos and smoking acting separately. If this test is significant and $\hat{\beta}_3$ is greater than 0, we would conclude that the joint effect of asbestos and smoking is greater than a multiplicative combination of separate effects. Or, if the test is significant and $\hat{\beta}_3$ is less than zero, we would conclude that the joint effect of asbestos and smoking is less than a multiplicative combination of separate effects.

IV. The *E, V, W* Model—A General Model Containing a (0, 1) Exposure and Potential Confounders and Effect Modifiers

The variables:
$E = (0, 1)$ exposure
$C_1, C_2, \cdots, C_p$ continuous or categorical

We are now ready to discuss a logistic model that considers the effects of several independent variables and, in particular, allows for the control of confounding and the assessment of interaction. We call this model the *E, V, W* model. We consider a single dichotomous (0, 1) exposure variable, denoted by E, and p extraneous variables C_1, C_2, and so on, up through C_p. The variables C_1 through C_p may be either continuous or categorical.

EXAMPLE

$$D = \text{CHD}_{(0, 1)}$$
$$E = \text{CAT}_{(0, 1)}$$
control variables
$$\begin{cases} C_1 = \text{AGE}_{\text{continuous}} \\ C_2 = \text{CHL}_{\text{continuous}} \\ C_3 = \text{SMK}_{(0, 1)} \\ C_4 = \text{ECG}_{(0, 1)} \\ C_5 = \text{HPT}_{(0, 1)} \end{cases}$$

As an example of this special case, suppose the disease variable is coronary heart disease status (CHD), the exposure variable E is catecholamine level (CAT); where 1 equals high and 0 equals low; and the control variables are AGE, cholesterol level (CHL), smoking status (SMK), electrocardiogram abnormality status (ECG), and hypertension status (HPT).

EXAMPLE (continued)

Model with eight independent variables:

$$\text{logit } P(\mathbf{X}) = \alpha + \beta \text{CAT}$$

$$\underbrace{+\gamma_1 \text{AGE} + \gamma_2 \text{CHL} + \gamma_3 \text{SMK} + \gamma_4 \text{ECG} + \gamma_5 \text{HPT}}_{\text{main effects}}$$

$$\underbrace{+\delta_1 \text{CAT} \times \text{CHL} + \delta_2 \text{CAT} \times \text{HPT}}_{\text{interaction effects}}$$

Parameters:
   α, β, γ's, and δ's instead of α and β's

where
   β: exposure variable
   γ's: potential confounders
   δ's: potential interaction variables

The general *E, V, W* Model

single exposure, controlling for
$C_1, C_2, ..., C_p$

We will assume here that both AGE and CHL are treated as continuous variables, that SMK is a (0, 1) variable, where 1 equals ever smoked and 0 equals never smoked, that ECG is a (0, 1) variable, where 1 equals abnormality present and 0 equals abnormality absent, and that HPT is a (0, 1) variable, where 1 equals high blood pressure and 0 equals normal blood pressure. There are, thus, five *C* variables in addition to the exposure variable CAT.

Corresponding to these variables is a model with eight independent variables. In addition to the exposure variable CAT, the model contains the five *C* variables as potential confounders plus two product terms involving two of the *C*'s, namely, CHL and HPT, which are each multiplied by the exposure variable CAT.

The model is written as logit P(**X**) equals α plus β times CAT plus the sum of five main effect terms γ_1 times AGE plus γ_2 times CHL and so on up through γ_5 times HPT plus the sum of δ_1 times CAT times CHL plus δ_2 times CAT times HPT. Here the five main effect terms account for the potential confounding effect of the variables AGE through HPT and the two product terms account for the potential interaction effects of CHL and HPT.

Note that the parameters in this model are denoted as α, β, γ's, and δ's, whereas previously we denoted all parameters other than the constant α as β_i's. We use β, γ's, and δ's here to distinguish different types of variables in the model. The parameter β indicates the coefficient of the exposure variable, the γ's indicate the coefficients of the potential confounders in the model, and the δ's indicate the coefficients of the potential interaction variables in the model. This notation for the parameters will be used throughout the remainder of this presentation.

Analogous to the above example, we now describe the general form of a logistic model, called the *E, V, W* model, that considers the effect of a single exposure controlling for the potential confounding and interaction effects of control variables C_1, C_2, up through C_p.

E, V, W Model

$k = p_1 + p_2 + 1 = $ # of variables in model
$p_1 = $ # of potential confounders
$p_2 = $ # of potential interactions
$1 = $ exposure variable

The general E, V, W model contains p_1 plus p_2 plus 1 variables, where p_1 is the number of potential confounders in the model, p_2 is the number of potential interaction terms in the model, and the 1 denotes the exposure variable.

CHD EXAMPLE

$p_1 = 5$: AGE, CHL, SMK, ECG, HPT
$p_2 = 2$: CAT × CHL, CAT × HPT
$p_1 + p_2 + 1 = 5 + 2 + 1 = 8$

In the CHD study example above, there are p_1 equal to five potential confounders, namely, the five control variables, and there is p_2 equal to two interaction variables, the first of which is CAT × CHL and the second is CAT × HPT. The total number of variables in the example is, therefore, p_1 plus p_2 plus 1 equals 5 plus 2 plus 1, which equals 8. This corresponds to the model presented earlier, which contained eight variables.

- $V_1, \cdots, V_{p_1}$ are potential confounders

- V's are functions of C's

In addition to the exposure variable E, the general model contains p_1 variables denoted as V_1, V_2 through V_{p_1}. The set of V's are functions of the C's that are thought to account for confounding in the data. We call the set of these V's **potential confounders.**

e.g., $V_1 = C_1$, $V_2 = (C_2)^2$, $V_3 = C_1 \times C_3$

For instance, we may have V_1 equal to C_1, V_2 equal to $(C_2)^2$, and V_3 equal to $C_1 \times C_3$.

CHD EXAMPLE

$V_1 = $ AGE, $V_2 = $ CHL, $V_3 = $ SMK,
$V_4 = $ ECG, $V_5 = $ HPT

The CHD example above has five V's that are the same as the C's.

- $W_1, \cdots, W_{p_2}$ are potential effect modifiers

- W's are functions of C's

e.g., $W_1 = C_1$, $W_2 = C_1 \times C_3$

Following the V's, we define p_2 variables which are product terms of the form E times W_1, E times W_2, and so on up through E times W_{p_2}, where W_1, W_2, through W_{p_2} denote a set of functions of the C's that are **potential effect modifiers with** E.

For instance, we may have W_1 equal to C_1 and W_2 equal to C_1 times C_3.

CHD EXAMPLE

$W_1 = $ CHL, $W_2 = $ HPT

The CHD example above has two W's, namely, CHL and HPT, that go into the model as product terms of the form CAT × CHL and CAT × HPT.

REFERENCES FOR CHOICE OF V's AND W's FROM C's

- Chapter 6: Modeling Strategy Guidelines
- *Epidemiologic Research*, Chapter 21

Assume: V's and W's are C's or subset of C's

EXAMPLE

$C_1 = \text{AGE}, C_2 = \text{RACE}, C_3 = \text{SEX}$
$V_1 = \text{AGE}, V_2 = \text{RACE}, V_3 = \text{SEX}$
$W_1 = \text{AGE}, W_2 = \text{SEX}$
$p_1 = 3, p_2 = 2, k = p_1 + p_2 + 1 = 6$

NOTE
W's ARE SUBSET OF V's

EXAMPLE

$\cancel{V_1 = \text{AGE}, V_2 = \text{RACE}}$
$\cancel{W_1 = \text{AGE}, W_2 = \text{SEX}}$

$$\text{logit P}(\mathbf{X}) = \alpha + \beta E + \gamma_1 V_1 + \gamma_2 V_2 + \cdots + \gamma_{p_1} V_{p_1}$$
$$+ \delta_1 EW_1 + \delta_2 EW_2 + \cdots + \delta_{p_2} EW_{p_2}$$

where
$\qquad \beta = \text{coefficient of } E$
$\qquad \gamma\text{'s} = \text{coefficient of } V\text{'s}$
$\qquad \delta\text{'s} = \text{coefficient of } W\text{'s}$

$$\text{logit P}(\mathbf{X}) = \alpha + \beta E$$
$$+ \sum_{i=1}^{p_1} \gamma_i V_i + E \sum_{j=1}^{p_2} \delta_j W_j$$

It is beyond the scope of this presentation to discuss the subtleties involved in the particular choice of the V's and W's from the C's for a given model. More depth is provided in a separate chapter (Chapter 6) on modeling strategies and in Chapter 21 of *Epidemiologic Research* by Kleinbaum, Kupper, and Morgenstern.

In most applications, the V's will be the C's themselves or some subset of the C's and the W's will also be the C's themselves or some subset thereof. For example, if the C's are AGE, RACE, and SEX, then the V's may be AGE, RACE, and SEX, and the W's may be AGE and SEX, the latter two variables being a subset of the C's. Here the number of V variables, p_1, equals 3, and the number of W variables, p_2, equals 2, so that k, which gives the total number of variables in the model, is p_1 plus p_2 plus 1 equals 6.

Note that although more details are given in the above references, you cannot have a W in the model that is not also contained in the model as a V; that is, W's have to be a subset of the V's. For instance, we cannot allow a model whose V's are AGE and RACE and whose W's are AGE and SEX because the SEX variable is not contained in the model as a V term.

A logistic model incorporating this special case containing the E, V, and W variables defined above can be written in logit form as shown here.

Note that β is the coefficient of the single exposure variable E, the γ's are coefficients of potential confounding variables denoted by the V's, and the δ's are coefficients of potential interaction effects involving E separately with each of the W's.

We can factor out the E from each of the interaction terms, so that the model may be more simply written as shown here. This is the form of the model that we will use henceforth in this presentation.

Adjusted odds ratio for $E = 1$ vs. $E = 0$
given $C_1, C_2, \cdots, C_p$ fixed

[handwritten: adjusted meaning other confounders kept fixed across exposure groups]

$$\text{ROR} = \exp\left(\beta + \sum_{j=1}^{p_2} \delta_j W_j\right)$$

- γ_i terms not in formula
- Formula assumes E is $(0, 1)$
- Formula is modified if E has other coding, e.g., $(1, -1)$, $(2, 1)$, ordinal, or interval
  (see Chapter 3 on coding)

[handwritten: This remains because E is multiple, and E varies over the two groups]

Interaction:
$$\text{ROR} = \exp\left(\beta + \sum \delta_j W_j\right)$$

- $\delta_j \neq 0 \Rightarrow \text{OR}$ depends on W_j
- Interaction $\Rightarrow$ effect of E differs at different levels of W's

[handwritten: e.g. interaction of smoking and asbestos exposure]

We now provide for this model an expression for an adjusted odds ratio that describes the effect of the exposure variable on disease status adjusted for the potential confounding and interaction effects of the control variables C_1 through C_p. That is, we give a formula for the risk odds ratio comparing the odds of disease development for exposed versus unexposed persons, with both groups having the same values for the extraneous factors C_1 through C_p. This formula is derived as a special case of the odds ratio formula for a general logistic model given earlier in our review.

For our special case, the odds ratio formula takes the form ROR equals e to the quantity β plus the sum from 1 through p_2 of the δ_j times W_j.

Note that β is the coefficient of the exposure variable E, that the δ_j are the coefficients of the interaction terms of the form E times W_j, and that the coefficients γ_i of the main effect variables V_i do not appear in the odds ratio formula.

Note also that this formula assumes that the dichotomous variable E is coded as a $(0, 1)$ variable with E equal to 1 for exposed persons and E equal to 0 for unexposed persons. If the coding scheme is different, for example, $(1, -1)$ or $(2, 1)$, or if E is an ordinal or interval variable, then the odds ratio formula needs to be modified. The effect of different coding schemes on the odds ratio formula will be described in Chapter 3.

This odds ratio formula tells us that if our model contains interaction terms, then the odds ratio will involve coefficients of these interaction terms and that, moreover, the value of the odds ratio will be different depending on the values of the W variables involved in the interaction terms as products with E. This property of the OR formula should make sense in that the concept of interaction implies that the effect of one variable, in this case E, is different at different levels of another variable, such as any of the W's.

- *V*'s not in OR formula but *V*'s in model, so OR formula controls confounding:

$$\text{logit P}(\mathbf{X}) = \alpha + \beta E + \Sigma \,\boxed{\gamma_i}\, V_i + E \,\Sigma \,\boxed{\delta_j}\, W_j$$

No interaction:

$$\text{all } \delta_j = 0 \Rightarrow \text{ROR} = \exp(\beta)$$
$$\uparrow$$
$$\text{constant}$$

$$\text{logit P}(\mathbf{X}) = \alpha + \beta E + \Sigma \,\gamma_i V_i$$
$$\uparrow$$
$$\text{confounding}$$
$$\text{effects adjusted}$$

EXAMPLE

The model:

$$\text{logit P}(\mathbf{X}) = \alpha + \beta \text{CAT}$$

$$\underbrace{+\gamma_1\text{AGE} + \gamma_2\text{CHL} + \gamma_3\text{SMK} + \gamma_4\text{ECG} + \gamma_5\text{HPT}}_{\text{main effects}}$$

$$\underbrace{+\text{CAT}(\delta_1\text{CHL} + \delta_2\text{HPT})}_{\text{interaction effects}}$$

$$\text{logit P}(\mathbf{X}) = \alpha + \beta \text{CAT}$$

$$\underbrace{+\gamma_1\text{AGE} + \gamma_2\text{CHL} + \gamma_3\text{SMK} + \gamma_4\text{ECG} + \gamma_5\text{HPT}}_{\text{main effects: confounding}}$$

$$\underbrace{+\text{CAT}(\delta_1\text{CHL} + \delta_2\text{HPT})}_{\text{product terms: interaction}}$$

$$\text{ROR} = \exp(\beta + \delta_1\text{CHL} + \delta_2\text{HPT})$$

Although the coefficients of the *V* terms do not appear in the odds ratio formula, these terms are still part of the fitted model. Thus, the odds ratio formula not only reflects the interaction effects in the model but also controls for the confounding variables in the model.

In contrast, if the model contains no interaction terms, then, equivalently, all the δ_j coefficients are 0; the odds ratio formula thus reduces to ROR equals to *e* to β, where β is the coefficient of the exposure variable *E*. Here, the **odds ratio is a fixed constant,** so that its value does not change with different values of the independent variables. The model in this case reduces to logit P(**X**) equals α plus β times *E* plus the sum of the main effect terms involving the *V*'s, and contains no product terms. For this model, we can say that *e* to β represents an odds ratio that **adjusts for the potential confounding effects** of the control variables C_1 through C_p defined in terms of the *V*'s.

As an example of the use of the odds ratio formula for the *E, V, W* model, we return to the CHD study example we described earlier. The CHD study model contained eight independent variables. The model is restated here as logit P(**X**) equals α plus β times CAT plus the sum of five main effect terms plus the sum of two interaction terms.

The five main effect terms in this model account for the potential confounding effects of the variables AGE through HPT. The two product terms account for the potential interaction effects of CHL and HPT.

For this example, the odds ratio formula reduces to the expression ROR equals *e* to the quantity β plus the sum δ_1 times CHL plus δ_2 times HPT.

EXAMPLE (continued)

$$ROR = \exp\left(\hat{\beta} + \hat{\delta}_1 CHL + \hat{\delta}_2 HPT\right)$$

- varies with values of CHL and HPT

AGE, SMK, and ECG are adjusted for confounding

n = 609 white males from Evans County, GA 9-year follow-up

Fitted model:

Variable	Coefficient
Intercept	$\hat{\alpha} = -4.0474$
CAT	$\hat{\beta} = -12.6809$
AGE	$\hat{\gamma}_1 = 0.0349$
CHL	$\hat{\gamma}_2 = 0.0055$
SMK	$\hat{\gamma}_3 = 0.7735$
ECG	$\hat{\gamma}_4 = 0.3665$
HPT	$\hat{\gamma}_5 = 1.0468$
CAT × CHL	$\hat{\delta}_1 = 0.0691$
CAT × HPT	$\hat{\delta}_2 = -2.3299$

$$\widehat{ROR} = \exp\left(-12.6809 + \underset{\text{exposure coefficient}}{0.0691 CHL} \underset{\text{interaction coefficient}}{-2.3299\ HPT}\right)$$

In using this formula, note that to obtain a numerical value for this odds ratio, not only do we need estimates of the coefficients β and the two δ's, but we also need to specify values for the variables CHL and HPT. In other words, once we have fitted the model to obtain estimates of the coefficients, we will get different values for the odds ratio depending on the values that we specify for the interaction variables in our model. Note, also, that although the variables AGE, SMK, and ECG are not contained in the odds ratio expression for this model, the confounding effects of these three variables plus CHL and HPT are being adjusted because the model being fit contains all five control variables as main effect V terms.

To provide numerical values for the above odds ratio, we will consider a data set of 609 white males from Evans County, Georgia, who were followed for 9 years to determine CHD status. The above model involving CAT, the five V variables, and the two W variables was fit to this data, and the fitted model is given by the list of coefficients corresponding to the variables listed here.

Based on the above fitted model, the estimated odds ratio for the CAT, CHD association adjusted for the five control variables is given by the expression shown here. Note that this expression involves only the coefficients of the exposure variable CAT and the interaction variables CAT times CHL and CAT times HPT, the latter two coefficients being denoted by δ's in the model.

EXAMPLE (continued)

$\widehat{ROR}$ varies with values of CHL and HPT

↙

interaction variables

- CHL = 220, HPT = 1

$\widehat{ROR} = \exp[-12.6809 + 0.0691(220) - 2.3299(1)]$
$= \exp(0.1912) = \boxed{1.21}$

- CHL = 200, HPT = 0

$\widehat{ROR} = \exp[-12.6809 + 0.0691(200) - 2.3299(0)]$
$= \exp(1.1391) = \boxed{3.12}$

CHL = 220, HPT = 1 $\Rightarrow \widehat{ROR} = 1.21$
CHL = 200, HPT = 0 $\Rightarrow \widehat{ROR} = 3.12$

controls for the confounding effects of AGE, CHL, SMK, ECG, and HPT

Choice of *W* values depends on investigator

EXAMPLE

TABLE OF POINT ESTIMATES $\widehat{ROR}$

	HPT = 0	HPT = 1
CHL = 180	0.78	0.08
CHL = 200	3.12	0.30
CHL = 220	12.44	1.21
CHL = 240	49.56	4.82

EXAMPLE

No interaction model for Evans County data ($n = 609$)

$\text{logit } P(\mathbf{X}) = \alpha + \beta\text{CAT}$

$\qquad + \gamma_1\text{AGE} + \gamma_2\text{CHL} + \gamma_3\text{SMK} + \gamma_4\text{ECG} + \gamma_5\text{HPT}$

This expression for the odds ratio tells us that we obtain a different value for the estimated odds ratio depending on the values specified for CHL and HPT. As previously mentioned, this should make sense conceptually because CHL and HPT are the only two interaction variables in the model, and by interaction, we mean that the odds ratio changes as the values of the interaction variables change.

To get a numerical value for the odds ratio, we consider, for example, the specific values CHL equal to 220 and HPT equal to 1. Plugging these into the odds ratio formula, we obtain *e* to the 0.1912, which equals 1.21.

As a second example, we consider CHL equal to 200 and HPT equal to 0. Here, the odds ratio becomes *e* to 1.1391, which equals 3.12.

Thus, we see that depending on the values of the interaction variables, we will get different values for the estimated odds ratios. Note that each estimated odds ratio obtained adjusts for the confounding effects of all five control variables because these five variables are contained in the fitted model as *V* variables.

In general, when faced with an odds ratio expression involving interaction (*W*) variables, the choice of values for the *W* variables depends primarily on the interest of the investigator. Typically, the investigator will choose a range of values for each interaction variable in the odds ratio formula; this choice will lead to a table of estimated odds ratios, such as the one presented here, for a range of CHL values and the two values of HPT. From such a table, together with a table of confidence intervals, the investigator can interpret the exposure–disease relationship.

As a second example, we consider a model containing no interaction terms from the same Evans County data set of 609 white males. The variables in the model are the exposure variable CAT, and five *V* variables, namely, AGE, CHL, SMK, ECG, and HPT. This model is written in logit form as shown here.

These are still RORs for exposure variable against disease, but now we must consider the interaction effects.

EXAMPLE (continued)

$\widehat{ROR} = \exp(\hat{\beta})$

Because this model contains no interaction terms, the odds ratio expression for the CAT, CHD association is given by e to the β "hat," where β "hat" is the estimated coefficient of the exposure variable CAT.

Fitted model:

When fitting this no interaction model to the data, we obtain estimates of the model coefficients that are listed here.

Variable	Coefficient
Intercept	$\hat{\alpha} = -6.7727$
CAT	$\hat{\beta} = 0.5976$
AGE	$\hat{\gamma}_1 = 0.0322$
CHL	$\hat{\gamma}_2 = 0.0087$
SMK	$\hat{\gamma}_3 = 0.8347$
ECG	$\hat{\gamma}_4 = 0.3695$
HPT	$\hat{\gamma}_5 = 0.4393$

$\widehat{ROR} = \exp(0.5976) = 1.82$

For this fitted model, then, the odds ratio is given by e to the power 0.5976, which equals 1.82. Note that this odds ratio is a fixed number, which should be expected, as there are no interaction terms in the model.

EXAMPLE COMPARISON

	interaction model	no interaction model
Intercept	−4.0474	−6.7727
CAT	−12.6809	0.5976
AGE	0.0349	0.0322
CHL	−0.0055	0.0087
SMK	0.7735	0.8347
ECG	0.3665	0.3695
HPT	1.0468	0.4393
CAT×CHL	0.0692	—
CAT×HPT	−2.3299	—

In comparing the results for the no interaction model just described with those for the model containing interaction terms, we see that the estimated coefficient for any variable contained in both models is different in each model. For instance, the coefficient of CAT in the **no interaction** model is 0.5976, whereas the coefficient of CAT in the **interaction** model is −12.6809. Similarly, the coefficient of AGE in the no interaction model is 0.0322, whereas the coefficient of AGE in the interaction model is 0.0349.

Which model? Requires *strategy*

It should not be surprising to see different values for corresponding coefficients as the two models give a different description of the underlying relationship among the variables. To decide which of these models, or maybe what other model, is more appropriate for this data, we need to use a **strategy** for model selection that includes carrying out tests of significance. A discussion of such a strategy is beyond the scope of this presentation but is described elsewhere (see Chapters 6 and 7).

V. Logistic Model for Matched Data (Chapter 8)

Focus: matched case-control studies

We will now consider a special case of the logistic model and the corresponding odds ratio for the **analysis of matched data.** This topic is discussed in more detail in Chapter 8. Our focus here will be on matched case-control studies, although the formulae provided also apply to matched follow-up studies.

Principle:
matched analysis $\Rightarrow$ stratified analysis

- strata are matched sets, e.g., pairs
  or
  combinations of matched sets, e.g., pooled pairs
- strata defined using dummy (indicator) variables

An important principle about modeling matched data is that a **matched analysis is a stratified analysis.** The strata are the matched sets, for example, the pairs in a matched pair design, or combinations of matched sets, such as pooled pairs within a close age range.

Moreover, when we use logistic regression to do a matched analysis, we define the strata using **dummy,** or indicator, variables.

$E = (0, 1)$ exposure

$C_1, C_2, \cdots, C_p$ control variables

- some C's matched by design
- remaining C's not matched

In defining a model for a matched analysis, we again consider the special case of a single $(0, 1)$ exposure variable of primary interest, together with a collection of control variables C_1, C_2, and so on up through C_p, to be adjusted in the analysis for possible confounding and interaction effects. We assume that some of these C variables have been matched in the study design, either by using pair matching, R-to-1 matching, or frequency matching. The remaining C variables have not been matched, but it is of interest to control for them, nevertheless.

$D = (0, 1)$ disease
$X_1 = E = (0, 1)$ exposure

Some X's: V_{1i} dummy variables
(matched status)

Some X's: V_{2i} variables
(potential confounders)

Some X's: product terms EW_j
(potential interaction variables)

$$\text{logit P}(\mathbf{X}) = \alpha + \beta E$$
$$+ \underbrace{\sum \gamma_{1i} V_{1i}}_{\text{matching}} + \underbrace{\sum \gamma_{2i} V_{2i}}_{\text{confounders}}$$
$$+ E \underbrace{\sum \delta_j W_j}_{\text{interaction}}$$

Given the above context, we will now define the following set of variables to be incorporated into a logistic model for matched data. We have a (0, 1) disease variable D and a (0, 1) exposure variable X_1 equal to E. We also have a collection of X's that are dummy variables indicating the different matched strata; these variables are denoted as V_{1i} variables.

Further, we have a collection of X's that are defined from the C's not involved in the matching. These X's represent potential confounders in addition to the matched variables and are denoted as V_{2i} variables.

Finally, we have a collection of X's which are product terms of the form E times W_j, where the W's denote potential effect modifiers. Note that the W's will usually be defined in terms of the V_2 variables.

The logistic model for a matched analysis is shown here. Note that the γ_{1i} are coefficients of the dummy variables for the matching strata, the γ_{2i} are the coefficients of the potential confounders not involved in the matching, and the δ_j are the coefficients of the interaction variables.

EXAMPLE

Pair matching by AGE, RACE, SEX
  100 matched pairs
    99 dummy variables

$$V_{1i} = \begin{cases} 1 & \text{if } i\text{th matched pair} \\ 0 & \text{otherwise} \end{cases}$$

$$i = 1, 2, \cdots, 99$$

$$V_{11} = \begin{cases} 1 & \text{if first matched pair} \\ 0 & \text{otherwise} \end{cases}$$

$$V_{12} = \begin{cases} 1 & \text{if second matched pair} \\ 0 & \text{otherwise} \end{cases}$$

$\vdots$

$$V_{1, 99} = \begin{cases} 1 & \text{if 99th matched pair} \\ 0 & \text{otherwise} \end{cases}$$

As an example of dummy variables defined for matched strata, consider a study involving pair matching by age, race, and sex, and containing 100 matched pairs. Then the above model requires defining 99 dummy variables to incorporate the 100 matched pairs. For example, we can define these variables as V_{1i} equals 1 if an individual falls into the ith matched pair and 0 otherwise. Thus, V_{11} equals 1 if an individual is in the first matched pair and 0 otherwise, V_{12} equals 1 if an individual is in the second matched pair and 0 otherwise, and so on up to $V_{1, 99}$, which equals 1 if an individual is in the 99th matched pair and 0 otherwise.

EXAMPLE (continued)

1st matched set:

$V_{11} = 1$, $V_{12} = V_{13} = \cdots = V_{1, 99} = 0$

99th matched set:

$V_{1, 99} = 1$, $V_{11} = V_{12} = \cdots = V_{1, 98} = 0$

100th matched set:

$V_{11} = V_{12} = \cdots = V_{1, 99} = 0$

Alternatively, when we use the above dummy variable definition, a person in the first matched set will have V_{11} equal to 1 and the remaining dummy variables equal to 0, a person in the 99th matched set will have $V_{1, 99}$ equal to 1 and the other dummy variables equal to 0, and a person in the last matched set will have all 99 dummy variables equal to 0.

$$\text{ROR} = \exp\left(\beta + \sum \delta_j W_j\right)$$

- OR formula for E, V, W model
- two types of V variables are controlled

The odds ratio formula for the matched analysis model is given by the expression ROR equals e to the quantity β plus the sum of the δ_j times the W_j. This is exactly the same odds ratio formula as given earlier for the E, V, W model for (0, 1) exposure variables. The matched analysis model is essentially an E, V, W model also, even though it contains two different types of V variables.

EXAMPLE

Case-control study
2-to-1 matching
$D = \text{MI}_{(0, 1)}$
$E = \text{SMK}_{(0, 1)}$

As an example of a matched pairs model, consider a case-control study using 2-to-1 matching that involves the following variables: The disease variable is myocardial infarction status (MI); the exposure variable is smoking status (SMK), a (0, 1) variable.

$\underbrace{C_1 = \text{AGE}, \; C_2 = \text{RACE}, \; C_3 = \text{SEX}, \; C_4 = \text{HOSPITAL}}_{\text{matched}}$

$\underbrace{C_5 = \text{SBP}, \; C_6 = \text{ECG}}_{\text{not matched}}$

There are six C variables to be controlled. The first four of these variables—age, race, sex, and hospital status—are involved in the matching, and the last two variables—systolic blood pressure (SBP) and electrocardiogram status (ECG)—are not involved in the matching.

$n = 117$ (39 matched sets)

The study involves 117 persons in 39 matched sets or strata, each strata containing 3 persons, 1 of whom is a case and the other 2 are matched controls.

$$\text{logit P}(\mathbf{X}) = \alpha + \beta \, \text{SMK} + \sum_{i=1}^{38} \gamma_{1i} V_{1i}$$

$$+ \gamma_{21} \text{SBP} + \gamma_{22} \text{ECG}$$

$$+ \text{SMK}(\delta_1 \text{SBP} + \delta_2 \text{ECG})$$

A logistic model for the above situation is shown here. This model contains 38 terms of the form γ_{1i} times V_{1i}, where V_{1i} are dummy variables for the 39 matched sets. The model also contains two potential confounders involving the two variables, SBP and ECG, not involved in the matching as well as two interaction variables involving these same two variables.

EXAMPLE (continued)

$$ROR = \exp\left(\beta + \delta_1 SBP + \delta_2 ECG\right)$$

- does not contain V's or γ's
- V's controlled as potential confounders

The odds ratio for the above logistic model is given by the formula e to the quantity β plus the sum of δ_1 times SBP and δ_2 times ECG. Note that this odds ratio expression does not contain any V terms or corresponding γ coefficients as such terms are potential confounders, not interaction variables. The V terms are nevertheless being controlled in the analysis because they are part of the logistic model being used.

This presentation is now complete. We have described important special cases of the logistic model, namely, models for

SUMMARY

1. Introduction
✓ 2. Important Special Cases

- simple analysis
- interaction assessment involving two variables
- assessment of potential confounding and interaction effects of several covariates
- matched analyses

3. Computing the Odds Ratio

We suggest that you review the material covered here by reading the detailed outline that follows. Then do the practice exercises and test.

All of the special cases in this presentation involved a (0, 1) exposure variable. In the next chapter, we consider how the odds ratio formula is modified for other codings of single exposures and also examine several exposure variables in the same model, controlling for potential confounders and effect modifiers.

Detailed Outline

I. **Overview** (page 42)

A. Focus:
- simple analysis
- multiplicative interaction
- controlling several confounders and effect modifiers
- matched data

B. Logistic model formula when $\mathbf{X} = (X_1, X_2, \ldots, X_k)$:

$$P(\mathbf{X}) = \frac{1}{1 + e^{-\left(\alpha + \sum\limits_{i=1}^{k} \beta_i X_i\right)}}.$$

C. Logit form of logistic model:

$$\text{logit } P(\mathbf{X}) = \alpha + \sum_{i=1}^{k} \beta_i X_i.$$

D. General odds ratio formula:

$$\text{ROR}_{\mathbf{X}_1, \mathbf{X}_0} = e^{\sum\limits_{i=1}^{k} \beta_i (X_{1i} - X_{0i})} = \prod_{i=1}^{k} e^{\beta_i (X_{1i} - X_{0i})}.$$

II. **Special case—Simple analysis** (pages 43–46)

A. The model:

$$P(\mathbf{X}) = \frac{1}{1 + e^{-(\alpha + \beta_1 E)}}$$

B. Logit form of the model:
$$\text{logit } P(\mathbf{X}) = \alpha + \beta_1 E$$

C. Odds ratio for the model: $\text{ROR} = \exp(\beta_1)$

D. Null hypothesis of no E, D effect: $H_0: \beta_1 = 0$.

E. The estimated odds ratio $\exp(\hat{\beta})$ is computationally equal to ad/bc where a, b, c, and d are the cell frequencies within the fourfold table for simple analysis.

III. **Assessing multiplicative interaction** (pages 46–52)

A. Definition of no interaction on a multiplicative scale:
$$\text{OR}_{11} = \text{OR}_{10} \times \text{OR}_{01},$$
where OR_{AB} denotes the odds ratio that compares a person in category A of one factor and category B of a second factor with a person in referent categories 0 of both factors, where A takes on the values 0 or 1 and B takes on the values 0 or 1.

B. Conceptual interpretation of no interaction formula: The effect of both variables A and B acting together is the same as the combined effect of each variable acting separately.

C. Examples of no interaction and interaction on a multiplicative scale.

D. A logistic model that allows for the assessment of multiplicative interaction:
$$\text{logit } P(\mathbf{X}) = \alpha + \beta_1 A + \beta_2 B + \beta_3 A \times B$$

E. The relationship of β_3 to the odds ratios in the no interaction formula above:
$$\beta_3 = \ln\left(\frac{OR_{11}}{OR_{10} \times OR_{01}}\right)$$

F. The null hypothesis of no interaction in the above two factor model: $H_0: \beta_3 = 0$.

IV. **The E, V, W model—A general model containing a (0, 1) exposure and potential confounders and effect modifiers** (pages 52–61)

A. Specification of variables in the model: start with E, C_1, C_2, . . ., C_p; then specify potential confounders V_1, V_2, . . ., V_{p_1}, which are functions of the C's, and potential interaction variables (i.e., effect modifiers) W_1, W_2, . . ., W_{p_2}, which are also functions of the C's and go into the model as product terms with E, i.e., $E \times W_i$.

B. The E, V, W model:
$$\text{logit } P(\mathbf{X}) = \alpha + \beta E + \sum_{i=1}^{p_1} \gamma_i V_i + E \sum_{j=1}^{p_2} \delta_j W_j$$

C. Odds ratio formula for the E, V, W model, where E is a (0, 1) variable:
$$ROR_{E=1 \text{ vs. } E=0} = \exp\left(\beta + \sum_{j=1}^{p_2} \delta_j W_j\right)$$

D. Odds ratio formula for E, V, W model if no interaction:
$ROR = \exp(\beta)$.

E. Examples of the E, V, W model: with interaction and without interaction

V. **Logistic model for matched data** (pages 61–64)

A. Important principle about matching and modeling: A matched analysis is a stratified analysis. The strata are the matched sets, e.g., pairs in a matched pairs analysis. Must use dummy variables to distinguish among matching strata in the model.

B. Specification of variables in the model:

i. Start with E and C_1, C_2, . . ., C_p.

ii. Then specify a collection of V variables that are dummy variables indicating the matching strata, i.e., V_{1i}, where i ranges from 1 to $G-1$, if there are G strata.

iii. Then specify V variables that correspond to potential confounders not involved in the matching (these are called V_{2i} variables).

iv. Finally, specify a collection of W variables that correspond to potential effect modifiers not involved in the matching (these are called W_j variables).

C. The logit form of a logistic model for matched data:

$$\text{logit } P(\mathbf{X}) = \alpha + \beta E + \sum_{i=1}^{G-1} \gamma_{1i} V_{1i} + \sum \gamma_{2i} V_{2i} + E \sum \delta_j W_j$$

D. The odds ratio expression for the above matched analysis model:

$$\text{ROR}_{E=1 \text{ vs. } E=0} = \exp\left(\beta + \sum \delta_j W_j\right)$$

E. The null hypothesis of no interaction in the above matched analysis model:

H_0: all $\delta_j = 0$.

F. Examples of a matched analysis model and odds ratio.

Practice Exercises

True or False (Circle T or F)

T F 1. A logistic model for a simple analysis involving a (0, 1) exposure variable is given by logit $P(\mathbf{X}) = \alpha + \beta E$, where E denotes the (0, 1) exposure variable.

T F 2. The odds ratio for the exposure–disease relationship in a logistic model for a simple analysis involving a (0, 1) exposure variable is given by β, where β is the coefficient of the exposure variable.

T F 3. The null hypothesis of no exposure–disease effect in a logistic model for a simple analysis is given by H_0: $\beta = 1$, where β is the coefficient of the exposure variable.

T F 4. The log of the estimated coefficient of a (0, 1) exposure variable in a logistic model for simple analysis is equal to ad/bc, where a, b, c, and d are the cell frequencies in the corresponding fourfold table for simple analysis.

T F 5. Given the model logit $P(\mathbf{X}) = \alpha + \beta E$, where E denotes a (0, 1) exposure variable, the **risk** for exposed persons ($E = 1$) is expressible as e^β.

T F 6. Given the model logit $P(\mathbf{X}) = \alpha + \beta E$, as in Exercise 5, the **odds** of getting the disease for exposed persons ($E = 1$) is given by $e^{\alpha+\beta}$.

T F   7. A logistic model that incorporates a multiplicative interaction effect involving two (0, 1) independent variables X_1 and X_2 is given by
logit $P(\mathbf{X}) = \alpha + \beta_1 X_1 + \beta_2 X_2 + \beta_3 X_1 X_2$.

T F   8. An equation that describes "no interaction on a multiplicative scale" is given by
$OR_{11} = OR_{10} / OR_{01}$.

T F   9. Given the model logit $P(\mathbf{X}) = \alpha + \beta E + \gamma SMK + \delta E \times SMK$, where E is a (0, 1) exposure variable and SMK is a (0, 1) variable for smoking status, the null hypothesis for a test of no interaction on a multiplicative scale is given by H_0: $\delta = 0$.

T F   10. For the model in Exercise 14, the odds ratio that describes the exposure disease effect controlling for smoking is given by $\exp(\beta + \delta)$.

T F   11. Given an exposure variable E and control variables AGE, SBP, and CHL, suppose it is of interest to fit a model that adjusts for the potential confounding effects of all three control variables considered as main effect terms and for the potential interaction effects with E of all three control variables. Then the logit form of a model that describes this situation is given by
logit $P(\mathbf{X}) = \alpha + \beta E + \gamma_1$ AGE $+ \gamma_2$ SBP $+ \gamma_3$ CHL $+ \delta_1$ AGE$\times$SBP $+ \delta_2$ AGE$\times$CHL $+ \delta_3$ SBP$\times$CHL.

T F   12. Given a logistic model of the form
logit $P(\mathbf{X}) = \alpha + \beta E + \gamma_1$ AGE $+ \gamma_2$ SBP $+ \gamma_3$ CHL,
where E is a (0, 1) exposure variable, the odds ratio for the effect of E adjusted for the confounding of AGE, CHL, and SBP is given by $\exp(\beta)$.

T F   13. If a logistic model contains interaction terms expressible as products of the form EW_j where W_j are potential effect modifiers, then the value of the odds ratio for the E, D relationship will be different, depending on the values specified for the W_j variables.

T F   14. Given the model logit $P(\mathbf{X}) = \alpha + \beta E + \gamma_1$ SMK $+ \gamma_2$ SBP, where E and SMK are (0, 1) variables, and SBP is continuous, then the odds ratio for estimating the effect of SMK on the disease, controlling for E and SBP is given by $\exp(\gamma_1)$.

T F   15. Given E, C_1, and C_2, and letting $V_1 = C_1 = W_1$ and $V_2 = C_2 = W_2$, then the corresponding logistic model is given by
logit $P(\mathbf{X}) = \alpha + \beta E + \gamma_1 C_1 + \gamma_2 C_2 + E(\delta_1 C_1 + \delta_2 C_2)$.

T F   16. For the model in Exercise 15, if $C_1 = 20$ and $C_2 = 5$, then the odds ratio for the E, D relationship has the form $\exp(\beta + 20\delta_1 + 5\delta_2)$.

Given a matched pairs case-control study with 100 subjects (50 pairs), suppose that in addition to the variables involved in the matching, the variable physical activity level (PAL) was measured but not involved in the matching.

T   F   17. For the matched pairs study described above, assuming no pooling of matched pairs into larger strata, a logistic model for a matched analysis that contains an intercept term requires 49 dummy variables to distinguish among the 50 matched pair strata.

T   F   18. For the matched pairs study above, a logistic model assessing the effect of a (0, 1) exposure E and controlling for the confounding effects of the matched variables and the unmatched variable PAL plus the interaction effect of PAL with E is given by the expression

$$\text{logit } P(\mathbf{X}) = \alpha + \beta E + \sum_{i=1}^{49} \gamma_1 V_i + \gamma_{50} \text{PAL} + \delta E \times \text{PAL},$$

where the V_i are dummy variables that indicate the matched pair strata.

T   F   19. Given the model in Exercise 18, the odds ratio for the exposure–disease relationship that controls for matching and for the confounding and interactive effect of PAL is given by $\exp(\beta + \delta\text{PAL})$.

T   F   20. Again, given the model in Exercise 18, the null hypothesis for a test of no interaction on a multiplicative scale can be stated as $H_0: \beta = 0$.

Test

True or False (Circle T or F)

T   F   1. Given the simple analysis model, logit $P(\mathbf{X}) = \phi + \psi Q$, where ϕ and ψ are unknown parameters and Q is a (0, 1) exposure variable, the odds ratio for describing the exposure–disease relationship is given by $\exp(\phi)$.

T   F   2. Given the model logit $P(\mathbf{X}) = \alpha + \beta E$, where E denotes a (0, 1) exposure variable, the **risk** for unexposed persons ($E = 0$) is expressible as $1/\exp(-\alpha)$.

T   F   3. Given the model in Question 2, the **odds** of getting the disease for unexposed persons ($E = 0$) is given by $\exp(\alpha)$.

T   F   4. Given the model logit $P(\mathbf{X}) = \phi + \psi\text{HPT} + \rho\text{ECG} + \pi\text{HPT}\times\text{ECG}$, where HPT is a (0, 1) exposure variable denoting hypertension status and ECG is a (0, 1) variable for electrocardiogram status, the null hypothesis for a test of no interaction on a multiplicative scale is given by $H_0: \exp(\pi) = 1$.

T F 5. For the model in Question 4, the odds ratio that describes the effect of HPT on disease status, controlling for ECG, is given by $\exp(\psi + \pi ECG)$.

T F 6. Given the model logit $P(\mathbf{X}) = \alpha + \beta E + \phi$ HPT $+ \psi$ ECG, where E, HPT, and ECG are (0, 1) variables, then the odds ratio for estimating the effect of ECG on the disease, controlling for E and HPT, is given by $\exp(\psi)$.

T F 7. Given E, C_1, and C_2, and letting $V_1 = C_1 = W_1$, $V_2 = (C_1)^2$, and $V_3 = C_2$, then the corresponding logistic model is given by
logit $P(\mathbf{X}) = \alpha + \beta E + \gamma_1 C_1 + \gamma_2 C_1^2 + \gamma_3 C_2 + \delta E C_1$.

T F 8. For the model in Question 7, if $C_1 = 5$ and $C_2 = 20$, then the odds ratio for the E, D relationship has the form $\exp(\beta + 20\delta)$.

Given a 4-to-1 case-control study with 100 subjects (i.e., 20 matched sets), suppose that in addition to the variables involved in the matching, the variables obesity (OBS) and parity (PAR) were measured but not involved in the matching.

T F 9. For the matched pairs study above, a logistic model assessing the effect of a (0, 1) exposure E, controlling for the confounding effects of the matched variables and the unmatched variables OBS and PAR plus the interaction effects of OBS with E and PAR with E, is given by the expression

$$\text{logit } P(\mathbf{X}) = \alpha + \beta E + \sum_{i=1}^{99} \gamma_{1i} V_{1i} + \gamma_{21} OBS + \gamma_{22} PAR + \delta_1 E \times OBS + \delta_2 E \times PAR.$$

T F 10. Given the model in Question 9, the odds ratio for the exposure–disease relationship that controls for matching and for the confounding and interactive effects of OBS and PAR is given by $\exp(\beta + \delta_1 E \times OBS + \delta_2 E \times PAR)$.

Consider a 1-year follow-up study of bisexual males to assess the relationship of behavioral risk factors to the acquisition of HIV infection. Study subjects were all in the 20 to 30 age range and were enrolled if they tested HIV negative and had claimed not to have engaged in "high-risk" sexual activity for at least 3 months. The outcome variable is HIV status at 1 year, a (0, 1) variable, where a subject gets the value 1 if HIV positive and 0 if HIV negative at 1 year after start of follow-up. Four risk factors were considered: consistent and correct condom use (CON), a (0, 1) variable; having one or more sex partners in high-risk groups (PAR), also a (0, 1) variable; the number of sexual partners (NP); and the average number of sexual contacts per month (ASCM). The primary purpose of this study was to determine the effectiveness of consistent and correct condom use in preventing the acquisition of HIV infection, controlling for the other variables. Thus, the variable CON is considered the exposure variable, and the variables PAR, NP, and ASCM are potential confounders and potential effect modifiers.

11. Within the above study framework, state the logit form of a logistic model for assessing the effect of CON on HIV acquisition, controlling for each of the other three risk factors as both potential confounders and potential effect modifiers. (Note: In defining your model, **only** use interaction terms that are two-way products of the form $E \times W$, where E is the exposure variable and W is an effect modifier.)

12. Using the model in Question 11, give an expression for the odds ratio that compares an exposed person (CON = 1) with an unexposed person (CON = 0) who has the same values for PAR, NP, and ASCM.

Suppose, instead of a follow-up study, that a matched pairs case-control study involving 200 pairs of bisexual males is performed, where the matching variables are NP and ASCM as described above, where PAR is also determined but is not involved in the matching, and where CON is the exposure variable.

13. Within the matched pairs case-control framework, state the logit form of a logistic model for assessing the effect of CON on HIV acquisition, controlling for PAR, NP, and ASCM as potential confounders and only PAR as an effect modifier.

14. Using the model in Question 13, give an expression for the risk of an exposed person (CON = 1) who is in the first matched pair and whose value for PAR is 1.

15. Using the model in Question 13, give an expression for the same odds ratio for the effect of CON, controlling for the confounding effects of NP, ASCM, and PAR and for the interaction effect of PAR.

Answers to Practice Exercises

1. T

2. F: OR = e^{β}

3. F: H_0: $\beta = 0$

4. F: $e^{\beta} = ad/bc$

5. F: risk for $E = 1$ is $1/[1 + e^{-(\alpha + \beta)}]$

6. T

7. T

8. F: $OR_{11} = OR_{10} \times OR_{01}$

9. T

10. F: $OR = \exp(\beta + \delta SMK)$

11. F: interaction terms should be $E \times AGE$, $E \times SBP$, and $E \times CHL$

12. T

13. T

14. T

15. T

16. T

17. T

18. T

19. T

20. F: $H_0: \delta = 0$

3

Computing the Odds Ratio in Logistic Regression

Contents

Introduction

In this chapter, the **E, V, W model** is extended to consider other coding schemes for a single exposure variable, including ordinal and interval exposures. The model is further extended to allow for several exposure variables. The formula for the odds ratio is provided for each extension, and examples are used to illustrate the formula.

Abbreviated Outline

The outline below gives the user a preview of the material covered by the presentation. Together with the objectives, this outline offers the user an overview of the content of this module. A detailed outline for review purposes follows the presentation.

Objectives   Upon completing this chapter, the learner should be able to:

1. Given a logistic model for a study situation involving a single exposure variable and several control variables, compute or recognize the expression for the odds ratio for the effect of exposure on disease status that adjusts for the confounding and interaction effects of functions of control variables:

    a. when the exposure variable is dichotomous and coded (*a*, *b*) for any two numbers *a* and *b*;

    b. when the exposure variable is ordinal and two exposure values are specified;

    c. when the exposure variable is continuous and two exposure values are specified.

2. Given a study situation involving a single nominal exposure variable with more than two (i.e., polytomous) categories, state or recognize a logistic model which allows for the assessment of the exposure–disease relationship controlling for potential confounding and assuming no interaction.

3. Given a study situation involving a single nominal exposure variable with more than two categories, compute or recognize the expression for the odds ratio that compares two categories of exposure status, controlling for the confounding effects of control variables and assuming no interaction.

4. Given a study situation involving several distinct exposure variables, state or recognize a logistic model that allows for the assessment of the joint effects of the exposure variables on disease controlling for the confounding effects of control variables and assuming no interaction.

5. Given a study situation involving several distinct exposure variables, state or recognize a logistic model that allows for the assessment of the joint effects of the exposure variables on disease controlling for the confounding and interaction effects of control variables.

Presentation

I. Overview

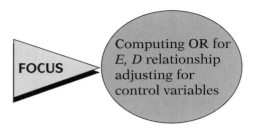

- dichotomous *E*—arbitrary coding
- ordinal or interval *E*
- polytomous *E*
- several *E*'s

This presentation describes how to compute the odds ratio for special cases of the general logistic model involving one or more exposure variables. We focus on models that allow for the assessment of an exposure–disease relationship that adjusts for the potential confounding and/or effect modifying effects of control variables.

In particular, we consider dichotomous exposure variables with arbitrary coding; that is, the coding of exposure may be other than (0, 1). We also consider single exposures which are ordinal or interval scaled variables. And, finally, we consider models involving several exposures, a special case of which involves a single polytomous exposure.

Chapter 2—*E, V, W* model:
- (0, 1) exposure
- confounders
- effect modifiers

In the previous chapter we described the logit form and odds ratio expression for the *E, V, W* logistic model, where we considered a single (0, 1) exposure variable and we allowed the model to control several potential confounders and effect modifiers.

The variables in the *E, V, W* model:

E: (0, 1) exposure

C's: control variables

V's: potential confounders

W's: potential effect modifiers (i.e., go into model as $E \times W$)

Recall that in defining the *E, V, W* model, we start with a single dichotomous (0, 1) exposure variable, *E*, and *p* control variables C_1, C_2, and so on, up through C_p. We then define a set of potential confounder variables, which are denoted as *V*'s. These *V*'s are functions of the *C*'s that are thought to account for confounding in the data. We then define a set of potential effect modifiers, which are denoted as *W*'s. Each of the *W*'s goes into the model as product term with *E*.

The *E, V, W* model:

$$\text{logit P}(\mathbf{X}) = \alpha + \beta E + \sum_{i=1}^{p_1} \gamma_i V_i + E \sum_{j=1}^{p_2} \delta_j W_j$$

The **logit form** of the *E, V, W* model is shown here. Note that β is the coefficient of the single exposure variable *E*, the gammas (γ's) are coefficients of potential confounding variables denoted by the *V*'s, and the deltas (δ's) are coefficients of potential interaction effects involving *E* separately with each of the *W*'s.

Adjusted odds ratio for effect of *E* adjusted for C's:

$$\text{ROR}_{E=1 \text{ vs. } E=0} = \exp\left(\beta + \sum_{j=1}^{p_2} \delta_j W_j\right)$$

(γ_i terms not in formula)

For this model, the formula for the **adjusted odds ratio** for the effect of the exposure variable on disease status adjusted for the potential confounding and interaction effects of the *C*'s is shown here. This formula takes the form ROR equals *e* to the quantity β plus the sum from terms of the form δ_j times W_j. Note that the coefficients γ_i of the main effect variables V_i do not appear in the odds ratio formula.

II. Odds Ratio for Other Codings of a Dichotomous *E*

Need to modify OR formula if coding of *E* is not (0, 1)

Note that this odds ratio formula assumes that the dichotomous variable *E* is coded as a (0, 1) variable with *E* equal to 1 when exposed and *E* equal to 0 when unexposed. If the coding scheme is different— for example, (−1, 1) or (2, 1), or if *E* is an ordinal or interval variable—then the odds ratio formula needs to be modified.

Focus: ✓ dichotomous
         ordinal
         interval

We now consider other coding schemes for dichotomous variables. Later, we also consider coding schemes for ordinal and interval variables.

$$E = \begin{cases} a & \text{if exposed} \\ b & \text{if unexposed} \end{cases}$$

$$\text{ROR}_{E=a \text{ vs. } E=b} = \exp\left[(a-b)\beta + (a-b)\sum_{j=1}^{p_2} \delta_j W_j\right]$$

Suppose *E* is coded to take on the value *a* if exposed and *b* if unexposed. Then, it follows from the general odds ratio formula that ROR equals *e* to the quantity (*a* − *b*) times β plus (*a* − *b*) times the sum of the δ_j times the W_j.

EXAMPLES

(A) $a = 1, b = 0 \Rightarrow (a-b) = (1-0) = 1$

$$\text{ROR} = \exp\left(1 \times \beta + 1 \times \sum \delta_j W_j\right)$$

(B) $a = 1, b = -1 \Rightarrow (a-b) = (1-[-1]) = 2$

$$\text{ROR} = \exp\left(2\beta + 2\sum \delta_j W_j\right)$$

(C) $a = 100, b = 0 \Rightarrow (a-b) = (100-0) = 100$

$$\text{ROR} = \exp\left(100\beta + 100\sum \delta_j W_j\right)$$

For example, if *a* equals 1 and *b* equals 0, then we are using the (0, 1) coding scheme described earlier. It follows that *a* minus *b* equals 1 minus 0, or 1, so that the ROR expression is *e* to the β plus the sum of the δ_j times the W_j. We have previously given this expression for (0, 1) coding.

In contrast, if *a* equals 1 and *b* equals −1, then *a* minus *b* equals 1 minus −1, which is 2, so the odds ratio expression changes to *e* to the quantity 2 times β plus 2 times the sum of the δ_j times the W_j.

As a third example, suppose *a* equals 100 and *b* equals 0, then *a* minus *b* equals 100, so the odds ratio expression changes to *e* to the quantity 100 times β plus 100 times the sum of the δ_j times the W_j.

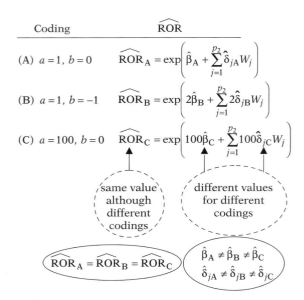

Coding	$\widehat{ROR}$
(A) $a = 1, b = 0$	$\widehat{ROR}_A = \exp\left(\hat{\beta}_A + \sum_{j=1}^{p_2} \hat{\delta}_{jA} W_j\right)$
(B) $a = 1, b = -1$	$\widehat{ROR}_B = \exp\left(2\hat{\beta}_B + \sum_{j=1}^{p_2} 2\hat{\delta}_{jB} W_j\right)$
(C) $a = 100, b = 0$	$\widehat{ROR}_C = \exp\left(100\hat{\beta}_C + \sum_{j=1}^{p_2} 100\hat{\delta}_{jC} W_j\right)$

same value although different codings

different values for different codings

$\widehat{ROR}_A = \widehat{ROR}_B = \widehat{ROR}_C$

$\hat{\beta}_A \neq \hat{\beta}_B \neq \hat{\beta}_C$
$\hat{\delta}_{jA} \neq \hat{\delta}_{jB} \neq \hat{\delta}_{jC}$

Thus, depending on the coding scheme for E, the odds ratio will be calculated differently. Nevertheless, even though β "hat" and the δ_j "hats" will be different for different coding schemes, the final odds ratio value will be the same as long as the correct formula is used for the corresponding coding scheme.

As shown here for the three examples above, which are labeled **A**, **B**, and **C**, the three computed odds ratios will be the same, even though the estimates β "hat" and δ_j "hat" used to compute these odds ratios will be different for different codings.

EXAMPLE: No Interaction Model

Evans County follow-up study:
  $n = 609$ white males
  $D = $ CHD status
  $E = $ CAT, dichotomous
  $V_1 = $ AGE, $V_2 = $ CHL, $V_3 = $ SMK,
  $V_4 = $ ECG, $V_5 = $ HPT

$\text{logit } P(\mathbf{X}) = \alpha + \beta\text{CAT}$

$\qquad + \gamma_1\text{AGE} + \gamma_2\text{CHL} + \gamma_3\text{SMK}$

$\qquad + \gamma_4\text{ECG} + \gamma_5\text{HPT}$

CAT: (0, 1) versus other codings

$\widehat{ROR} = \exp(\hat{\beta})$

As a numerical example, we consider a model that contains no interaction terms from a data set of 609 white males from Evans County, Georgia. The study is a follow-up study to determine the development of coronary heart disease (CHD) over 9 years of follow-up. The variables in the model are CAT, a dichotomous exposure variable, and five V variables, namely, AGE, CHL, SMK, ECG, and HPT.

This model is written in **logit form** as logit $P(\mathbf{X})$ equals α plus β times CAT plus the sum of five main effect terms γ_1 times AGE plus γ_2 times CHL, and so on up through γ_5 times HPT.

We first describe the results from fitting this model when CAT is coded as a (0, 1) variable. Then, we contrast these results with other codings of CAT.

Because this model contains no interaction terms and CAT is coded as (0, 1), the odds ratio expression for the CAT, CHD association is given by e to β "hat," where β "hat" is the estimated coefficient of the exposure variable CAT.

EXAMPLE (continued)

(0, 1) coding for CAT

Variable	Coefficient
Intercept	$\hat{\alpha} = -6.7727$
CAT	$\hat{\beta} = 0.5976$
AGE	$\hat{\gamma}_1 = 0.0322$
CHL	$\hat{\gamma}_2 = 0.0087$
SMK	$\hat{\gamma}_3 = 0.8347$
ECG	$\hat{\gamma}_4 = 0.3695$
HPT	$\hat{\gamma}_5 = 0.4393$

$\widehat{ROR} = \exp(0.5976) = \boxed{1.82}$

No interaction model: ROR fixed

$(-1, 1)$ coding for CAT : $\hat{\beta} = 0.2988 = \left(\dfrac{0.5976}{2}\right)$

$$\widehat{ROR} = \exp(2\hat{\beta}) = \exp(2 \times 0.2988)$$
$$= \exp(0.5976)$$
$$= 1.82$$

same $\widehat{ROR}$ as for (0, 1) coding

Note: $\widehat{ROR} \neq \exp(0.2988) = \underset{\underset{\text{incorrect value}}{\uparrow}}{1.35}$

Fitting this no interaction model to the data, we obtain the estimates listed here.

For this fitted model, then, the odds ratio is given by *e* to the power 0.5976, which equals 1.82. Notice that, as should be expected, this odds ratio is a fixed number as there are no interaction terms in the model.

Now, if we consider the same data set and the same model, except that the coding of CAT is $(-1, 1)$ instead of (0, 1), the coefficient β "hat" of CAT becomes 0.2988, which is one-half of 0.5976. Thus, for this coding scheme, the odds ratio is computed as *e* to 2 times the corresponding β "hat" of 0.2988, which is the same as *e* to 0.5976, or 1.82. We see that, regardless of the coding scheme used, the final odds ratio result is the same, as long as the correct odds ratio formula is used. In contrast, it would be incorrect to use the $(-1, 1)$ coding scheme and then compute the odds ratio as *e* to 0.2988.

III. Odds Ratio for Arbitrary Coding of *E*

Model:

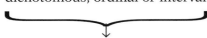
dichotomous, ordinal or interval

$$\text{logit P}(\mathbf{X}) = \alpha + \beta E + \sum_{i=1}^{p_1} \gamma_i V_i + E \sum_{j=1}^{p_2} \delta_j W_j$$

We now consider the odds ratio formula for any single exposure variable *E*, whether **dichotomous, ordinal,** or **interval,** controlling for a collection of *C* variables in the context of an *E, V, W* model shown again here. That is, we allow the variable *E* to be defined arbitrarily of interest.

E^* (group 1) vs. E^{**} (group 2)

To obtain an odds ratio for such a generally defined E, we need to specify two values of E to be compared. We denote the two values of interest as E^* and E^{**}. We need to specify two values because an odds ratio requires the **comparison of two groups**—in this case two levels of the exposure variable E—even when the exposure variable can take on more than two values, as when E is ordinal or interval.

$$\text{ROR}_{E^* \text{ vs. } E^{**}} = \exp\left[(E^*-E^{**})\beta+(E^*-E^{**})\sum_{j=1}^{p_2}\delta_j W_j\right]$$

Same as

$$\text{ROR}_{E=a \text{ vs. } E=b} = \exp\left[(a-b)\beta+(a-b)\sum_{j=1}^{p_2}\delta_j W_j\right]$$

The odds ratio formula for E^* versus E^{**}, equals e to the quantity $(E^* - E^{**})$ times β plus $(E^* - E^{**})$ times the sum of the δ_j times W_j. This is essentially the same formula as previously given for dichotomous E, except that here, several different odds ratios can be computed as the choice of E^* and E^{**} ranges over the possible values of E.

EXAMPLE

$E = \text{SSU} = \text{social support status } (0\text{–}5)$

We illustrate this formula with several examples. First, suppose E gives social support status as denoted by SSU, which is an index ranging from 0 to 5, where 0 denotes a person without any social support and 5 denotes a person with the maximum social support possible.

(A) $\text{SSU}^* = 5$ vs. $\text{SSU}^{**} = 0$

$$\text{ROR}_{5,0} = \exp\,[(\text{SSU}^* - \text{SSU}^{**})$$
$$\beta + (\text{SSU}^* - \text{SSU}^{**})\,\Sigma\,\delta_j\,W_j]$$

$$= \exp\,[(5-0)\,\beta + (5-0)\Sigma\,\delta_j\,W_j]$$

$$= \exp\,(5\beta + 5\Sigma\,\delta_j\,W_j)$$

To obtain an odds ratio involving **social support status (SSU),** in the context of our E, V, W model, we need to specify two values of E. One such pair of values is SSU* equals 5 and SSU** equals 0, which compares the odds for persons who have the highest amount of social support with the odds for persons who have the lowest amount of social support. For this choice, the odds ratio expression becomes e to the quantity $(5 - 0)$ times β plus $(5 - 0)$ times the sum of the δ_j times W_j, which simplifies to e to 5β plus 5 times the sum of the δ_j times W_j.

(B) $\text{SSU}^* = 3$ vs. $\text{SSU}^{**} = 1$

$$\text{ROR}_{3,\,1} = \exp\left[(3-1)\beta+(3-1)\sum\delta_j W_j\right]$$

$$= \exp\left(2\beta+2\sum\delta_j W_j\right)$$

Similarly, if SSU* equals 3 and SSU** equals 1, then the odds ratio becomes e to the quantity $(3 - 1)$ times β plus $(3 - 1)$ times the sum of the δ_j times W_j, which simplifies to e to 2β plus 2 times the sum of the δ_j times W_j.

EXAMPLE (continued)

(C) SSU* = 4 vs. SSU** = 2

$$ROR_{4,2} = \exp\left[(4-2)\beta + (4-2)\sum \delta_j W_j\right]$$
$$= \exp\left(2\beta + 2\sum \delta_j W_j\right)$$

Note: ROR depends on the difference
$(E^* - E^{**})$, e.g., $(3 - 1) = (4 - 2) = 2$

EXAMPLE

E = SBP = systolic blood pressure
(interval)

(A) SBP* = 160 vs. SBP** = 120

$$ROR_{160, 120} = \exp\left[(SBP^* - SBP^{**})\beta + (SBP^* - SBP^{**})\sum \delta_j W_j\right]$$
$$= \exp\left[(160 - 120)\beta + (160 - 120)\sum \delta_j W_j\right]$$
$$= \exp\left(40\beta + 40\sum \delta_j W_j\right)$$

(B) SBP* = 200 vs. SBP** = 120

$$ROR_{200, 120} = \exp\left[(200 - 120)\beta + (200 - 120)\sum \delta_j W_j\right]$$
$$= \exp\left(80\beta + 80\sum \delta_j W_j\right)$$

No interaction:
$$ROR_{E^* \text{ vs. } E^{**}} = \exp\left[(E^* - E^{**})\beta\right]$$
If $(E^* - E^{**}) = 1$, then ROR $= \exp(\beta)$
e.g., $E^* = 1$ vs. $E^{**} = 0$
or $E^* = 2$ vs. $E^{**} = 1$

EXAMPLE

E = SBP
ROR $= \exp(\beta) \Rightarrow (SBP^* - SBP^{**}) = 1$
$\qquad\qquad\qquad\qquad$ not interesting$\uparrow$

Choice of SBP:
  Clinically meaningful categories,
  e.g., SBP* = 160, SBP* = 120

Strategy: Use quintiles of SBP

Quintile #	1	2	3	4	5
Mean or median	120	140	160	180	200

Note that if SSU* equals 4 and SSU** equals 2, then the odds ratio expression becomes 2β plus 2 times the sum of the δ_j times W_j, which is the same expression as obtained when SSU* equals 3 and SSU** equals 1. This occurs because the odds ratio depends on the difference between E^* and E^{**}, which in this case is 2, regardless of the specific values of E^* and E^{**}.

As another illustration, suppose E is the interval variable systolic blood pressure denoted by SBP. Again, to obtain an odds ratio, we must specify two values of E to compare. For instance, if SBP* equals 160 and SBP** equals 120, then the odds ratio expression becomes ROR equals e to the quantity $(160 - 120)$ times β plus $(160 - 120)$ times the sum of the δ_j times W_j, which simplifies to 40 times β plus 40 times the sum of the δ_j times W_j.

Or if SBP* equals 200 and SBP** equals 120, then the odds ratio expression becomes ROR equals e to the 80 times β plus 80 times the sum of the γ_j times W_j.

Note that in the no interaction case, the odds ratio formula for a general exposure variable E reduces to e to the quantity $(E^* - E^{**})$ times β. This is not equal to e to the β unless the difference $(E^* - E^{**})$ equals 1, as, for example, if E^* equals 1 and E^{**} equals 0, or E^* equals 2 and E^{**} equals 1.

Thus, if E denotes SBP, then the quantity e to β gives the odds ratio for comparing any two groups which differ by one unit of SBP. A one unit difference in SBP is not typically of interest, however. Rather, a typical choice of SBP values to be compared represent clinically meaningful categories of blood pressure, as previously illustrated, for example, by SBP* equals 160 and SBP** equals 120.

One possible strategy for choosing values of SBP* and SBP** is to categorize the distribution of SBP values in our data into clinically meaningful categories, say, quintiles. Then, using the mean or median SBP in each quintile, we can compute odds ratios comparing all possible pairs of mean or median SBP values.

EXAMPLE (continued)

SBP*	SBP**	OR
200	120	✓
200	140	✓
200	160	✓
200	180	✓
180	120	✓
180	140	✓
180	160	✓
160	140	✓
160	120	✓
140	120	✓

For instance, suppose the medians of each quintile are 120, 140, 160, 180, and 200. Then odds ratios can be computed comparing SBP* equal to 200 with SBP** equal to 120, followed by comparing SBP* equal to 200 with SBP** equal to 140, and so on until all possible pairs of odds ratios are computed. We would then have a table of odds ratios to consider for assessing the relationship of SBP to the disease outcome variable. The check marks in the table shown here indicate pairs of odds ratios that compare values of SBP* and SBP**.

IV. The Model and Odds Ratio for a Nominal Exposure Variable (No Interaction Case)

Several exposures: $E_1, E_2, \ldots, E_q$

- model
- odds ratio

The final special case of the logistic model that we will consider expands the E, V, W model to allow for several exposure variables. That is, instead of having a single E in the model, we will allow several E's, which we denote by E_1, E_2, and so on up through E_q. In describing such a model, we consider some examples and then give a general model formula and a general expression for the odds ratio.

Nominal variable: > 2 categories

e.g., ✓ occupational status in four groups

~~SSU (0 = 5) ordinal~~

First, suppose we have a single nominal exposure variable of interest; that is, instead of being dichotomous, the exposure contains more than two categories that are not orderable. An example is a variable such as occupational status, which is denoted in general as OCC, but divided into four groupings or occupational types. In contrast, a variable like social support, which we previously denoted as SSU and takes on discrete values ordered from 0 to 5, is an ordinal variable.

k categories $\Rightarrow k - 1$ dummy variables
$E_1, E_2, \ldots, E_{k-1}$

When considering nominal variables in a logistic model, we use dummy variables to distinguish the different categories of the variable. If the model contains an intercept term α, then we use $k-1$ dummy variables E_1, E_2, and so on up to E_{k-1} to distinguish among k categories.

EXAMPLE

$E = $ OCC with $k = 4 \Rightarrow k - 1 = 3$
$$\text{OCC}_1, \text{OCC}_2,$$
$$\text{OCC}_3$$

where $\text{OCC}_i = \begin{cases} 1 & \text{if category } i \\ 0 & \text{if otherwise} \end{cases}$

for $i = 1, 2, 3$

So, for example, with occupational status, we define three dummy variables OCC_1, OCC_2, and OCC_3 to reflect four occupational categories, where OCC_i is defined to take on the value 1 for a person in the ith occupational category and 0 otherwise, for i ranging from 1 to 3.

No interaction model:
$$\text{logit P}(\mathbf{X}) = \alpha + \beta_1 E_1 + \beta_2 E_2 + \ldots$$
$$+ \beta_{k-1} E_{k-1} + \sum_{i=1}^{p_1} \gamma_i V_i$$

A no interaction model for a nominal exposure variable with k categories then takes the form logit P($\mathbf{X}$) equals α plus β_1 times E_1 plus β_2 times E_2 and so on up to β_{k-1} times E_{k-1} plus the usual set of V terms, where the E_i are the dummy variables described above.

$$\text{logit P}(\mathbf{X}) = \alpha + \beta_1 \text{OCC}_1 + \beta_2 \text{OCC}_2$$
$$+ \beta_3 \text{OCC}_3 + \sum_{i=1}^{p_1} \gamma_i V_i$$

The corresponding model for four occupational status categories then becomes logit P($\mathbf{X}$) equals α plus β_1 times OCC_1 plus β_2 times OCC_2 plus β_3 times OCC_3 plus the V terms.

Specify $\mathbf{E}^*$ and $\mathbf{E}^{**}$ in terms of $k - 1$ dummy variables where

$$\mathbf{E} = (E_1, E_2, \ldots, E_{k-1})$$

To obtain an odds ratio from the above model, we need to specify two categories $\mathbf{E}^*$ and $\mathbf{E}^{**}$ of the nominal exposure variable to be compared, and we need to define these categories in terms of the $k - 1$ dummy variables. Note that we have used **bold letters** to **identify the two categories of** E; this has been done because the E variable is a collection of dummy variables rather than a single variable.

EXAMPLE

$E = $ occupational status (four categories)

$\mathbf{E}^* = $ category 3 vs. $\mathbf{E}^{**} = $ category 1

$\mathbf{E}^* = (\text{OCC}_1{}^* = 0, \text{OCC}_2{}^* = 0, \text{OCC}_3{}^* = 1)$

$\mathbf{E}^{**} = (\text{OCC}_1{}^{**} = 1, \text{OCC}_2{}^{**} = 0, \text{OCC}_3{}^{**} = 0)$

For the occupational status example, suppose we want an odds ratio comparing occupational category 3 with occupational category 1. Here, $\mathbf{E}^*$ represents category 3 and $\mathbf{E}^{**}$ represents category 1. In terms of the three dummy variables for occupational status, then, $\mathbf{E}^*$ is defined by $\text{OCC}_1{}^* = 0$, $\text{OCC}_2{}^* = 0$, and $\text{OCC}_3{}^* = 1$, whereas $\mathbf{E}^{**}$ is defined by $\text{OCC}_1{}^{**} = 1$, $\text{OCC}_2{}^{**} = 0$, and $\text{OCC}_3{}^{**} = 0$.

Generally, define $\mathbf{E}^*$ and $\mathbf{E}^{**}$ as
$$\mathbf{E}^* = (E_1{}^*, E_2{}^*, \ldots, E_{k-1}{}^*)$$
and
$$\mathbf{E}^* = (E_1{}^{**}, E_2{}^{**}, \ldots, E_{k-1}{}^{**})$$

More generally, category $\mathbf{E}^*$ is defined by the dummy variable values $E_1{}^*$, $E_2{}^*$, and so on up to $E_{k-1}{}^*$, which are 0's or 1's. Similarly, category $E_1{}^{**}$ is defined by the values $E_1{}^{**}$, $E_2{}^{**}$, and so on up to $E_{k-1}{}^{**}$, which is a different specification of 0's or 1's.

No interaction model

No interaction model

$$\text{ROR}_{E^* \text{ vs. } E^{**}}$$

$$= \exp\left[(E_1^* - E_1^{**})\,\beta_1 + (E_2^* - E_2^{**})\beta_2 \right.$$
$$\left. + \ldots + (E_{k-1}^* - E_{k-1}^{**})\beta_{k-1}\right]$$

EXAMPLE (OCC)

$$\text{ROR}_{3 \text{ vs. } 1} = \exp\left[(\overset{0}{\text{OCC}_1^*} - \overset{1}{\text{OCC}_1^{**}})\beta_1 \right.$$
$$+ (\overset{0}{\text{OCC}_2^*} - \overset{0}{\text{OCC}_2^{**}})\beta_2$$
$$\left. + (\overset{1}{\text{OCC}_3^*} - \overset{0}{\text{OCC}_3^{**}})\beta_3\right]$$

$$= \exp\left[(0 - 1)\beta_1 + (0 - 0)\beta_2 \right.$$
$$\left. + (1 - 0)\beta_3\right]$$

$$= \exp\left[(-1)\beta_1 + (0)\beta_2 + (1)\beta_3\right]$$

$$= \exp(-\beta_1 + \beta_3)$$

$$\widehat{\text{ROR}} = \exp(-\hat{\beta}_1 + \hat{\beta}_3)$$

E^* = category 3 vs. E^{**} = category 2:

$E^* = (\text{OCC}_1^* = 0, \text{OCC}_2^* = 0, \text{OCC}_3^* = 1)$

$E^{**} = (\text{OCC}_1^{**} = 0, \text{OCC}_2^{**} = 1, \text{OCC}_3^{**} = 0)$

$$\text{ROR}_{3 \text{ vs. } 2} = \exp\left[(0 - 0)\beta_1 + (0 - 1)\beta_2 \right.$$
$$\left. + (1 - 0)\beta_3\right]$$

$$= \exp\left[(0)\beta_1 + (-1)\beta_2 + (1)\beta_3\right]$$

$$= \exp(-\beta_2 + \beta_3)$$

Note: $\text{ROR}_{3 \text{ vs. } 1} = \exp(-\beta_1 + \beta_3)$

The **general odds ratio formula** for comparing two categories, E^* versus E^{**} of a general nominal exposure variable in a **no interaction logistic model**, is given by the formula ROR equals e to the quantity $(E_1^* - E_1^{**})$ times β_1 plus $(E_2^* - E_2^{**})$ times β_2, and so on up to $(E_{k-1}^* - E_{k-1}^{**})$ times β_{k-1}. When applied to a specific situation, this formula will usually involve more than one β_i in the exponent.

For example, when comparing occupational status category 3 with category 1, the odds ratio formula is computed as e to the quantity $(\text{OCC}_1^* - \text{OCC}_1^{**})$ times β_1 plus $(\text{OCC}_2^* - \text{OCC}_2^{**})$ times β_2 plus $(\text{OCC}_3^* - \text{OCC}_3^{**})$ times β_3.

When we plug in the values for OCC* and OCC**, this expression equals e to the quantity $(0 - 1)$ times β_1 plus $(0 - 0)$ times β_2 plus $(1 - 0)$ times β_3, which equals e to -1 times β_1 plus 0 times β_2 plus 1 times β_3, which reduces to e to the quantity $(-\beta_1)$ plus β_3.

We can obtain a single value for the estimate of this odds ratio by fitting the model and replacing β_1 and β_3 with their corresponding estimates β_1 "hat" and β_3 "hat." Thus, ROR "hat" for this example is given by e to the quantity $(-\beta_1$ "hat") plus β_3 "hat."

In contrast, if category 3 is compared to category 2, then E^* takes on the values 0, 0, and 1 as before, whereas E^{**} is now defined by $\text{OCC}_1^{**} = 0$, $\text{OCC}_2^{**} = 1$, and $\text{OCC}_3^{**} = 0$.

The odds ratio is then computed as e to the $(0 - 0)$ times β_1 plus $(0 - 1)$ times β_2 plus $(1 - 0)$ times β_3, which equals e to the 0 times β_1 plus -1 times β_2 plus 1 times β_3, which reduces to e to the quantity $(-\beta_2)$ plus β_3.

This odds ratio expression involves β_2 and β_3, whereas the previous odds ratio expression that compared category 3 with category 1 involved β_1 and β_3.

V. The Model and Odds Ratio for Several Exposure Variables (No Interaction Case)

q variables: $E_1, E_2, \ldots, E_q$

(dichotomous, ordinal, or interval)

We now consider the odds ratio formula when there are several different exposure variables in the model, rather than a single exposure variable with several categories. The formula for this situation is actually no different than for a single nominal variable. The different exposure variables may be denoted by E_1, E_2, and so on up through E_q. However, rather than being dummy variables, these E's can be any kind of variable—dichotomous, ordinal, or interval.

EXAMPLE

$E_1 = $ SMK (0,1)

$E_2 = $ PAL (ordinal)

$E_3 = $ SBP (interval)

For example, E_1 may be a (0, 1) variable for smoking (SMK), E_2 may be an ordinal variable for physical activity level (PAL), and E_3 may be the interval variable systolic blood pressure (SBP).

No interaction model:

$$\text{logit } P(\mathbf{X}) = \alpha + \beta_1 E_1 + \beta_2 E_2 + \ldots + \beta_q E_q$$
$$+ \sum_{i=1}^{p_1} \gamma_i V_i$$

- $q \neq k - 1$ in general

A no interaction model with several exposure variables then takes the form logit $P(\mathbf{X})$ equals α plus β_1 times E_1 plus β_2 times E_2, and so on up to β_q times E_q plus the usual set of V terms. This model form is the same as that for a single nominal exposure variable, although this time there are q E's of any type, whereas previously we had $k - 1$ dummy variables to indicate k exposure categories. The corresponding model involving the three exposure variables SMK, PAL, and SBP is shown here.

EXAMPLE

$$\text{logit } P(\mathbf{X}) = \alpha + \beta_1 \text{SMK} + \beta_2 \text{PAL} + \beta_3 \text{SBP}$$
$$+ \sum_{i=1}^{p_1} \gamma_i V_i$$

$\mathbf{E}^*$ vs. $\mathbf{E}^{**}$

$\mathbf{E}^* = (E_1^*, E_2^*, \ldots, E_q^*)$

$\mathbf{E}^{**} = (E_1^{**}, E_2^{**}, \ldots, E_q^{**})$

As before, the general odds ratio formula for several variables requires specifying the values of the exposure variables for two different persons or groups to be compared—denoted by the bold $\mathbf{E}^*$ and $\mathbf{E}^{**}$. Category $\mathbf{E}^*$ is specified by the variable values E_1^*, E_2^*, and so on up to E_q^*, and category $\mathbf{E}^{**}$ is specified by a different collection of values E_1^{**}, E_2^{**}, and so on up to E_q^{**}.

$$\text{ROR}_{\mathbf{E}^* \text{ vs. } \mathbf{E}^{**}} = \exp\left[\left(E_1^* - E_1^{**}\right)\beta_1\right.$$
$$+ \left(E_2^* - E_2^{**}\right)\beta_2 + \cdots$$
$$\left. + \left(E_q^* - E_q^{**}\right)\beta_q\right]$$

The general odds ratio formula for comparing $\mathbf{E}^*$ versus $\mathbf{E}^{**}$ is given by the formula ROR equals $\mathbf{e}$ to the quantity $(E_1^* - E_1^*)$ times β_1 plus $(E^* - E^{**})$ times β_2, and so on up to $(E_q^* - E_q^{**})$ times β_q.

In general

- q variables $\neq k - 1$ dummy variables

This formula is the same as that for a single exposure variable with several categories, except that here we have q variables, whereas previously we had $k - 1$ dummy variables.

EXAMPLE

$$\text{logit } P(X) = \alpha + \beta_1 \text{ SMK} + \beta_2 \text{ PAL} + \beta_3 \text{ SBP} + \gamma_1 \text{ AGE} + \gamma_2 \text{ SEX}$$

Nonsmoker, PAL = 25, SBP = 160

vs.

Smoker, PAL = 10, SBP = 120

$\mathbf{E^*} = (\text{SMK}^*=0, \text{PAL}^*=25, \text{SBP}^*=160)$

$\mathbf{E^{**}} = (\text{SMK}^{**}=1, \text{PAL}^{**}=10, \text{SBP}^{**}=120)$

AGE and SEX fixed, but unspecified

$$\text{ROR}_{\mathbf{E^* \text{ vs. } E^{**}}} = \exp\big[(\text{SMK}^* - \text{SMK}^{**})\beta_1$$
$$+ (\text{PAL}^* - \text{PAL}^{**})\beta_2$$
$$+ (\text{SBP}^* - \text{SBP}^{**})\beta_3\big]$$

$$= \exp\big[(0-1)\beta_1 + (25-10)\beta_2 + (160-120)\beta_3\big]$$

$$= \exp\big[(-1)\beta_1 + (15)\beta_2 + (40)\beta_3\big]$$

$$= \exp\big(-\beta_1 + 15\beta_2 + 40\beta_3\big)$$

$$\widehat{\text{ROR}} = \exp\big(-\hat{\beta}_1 + 15\hat{\beta}_2 + 40\hat{\beta}_3\big)$$

As an example, consider the three exposure variables defined above—SMK, PAL, and SBP. The control variables are AGE and SEX, which are defined in the model as V terms.

Suppose we wish to compare a nonsmoker who has a PAL score of 25 and systolic blood pressure of 160 to a smoker who has a PAL score of 10 and systolic blood pressure of 120, controlling for AGE and SEX. Then, here, $\mathbf{E^*}$ is defined by SMK*=0, PAL*=25, and SBP*=160, whereas $\mathbf{E^{**}}$ is defined by SMK**=1, PAL**=10, and SBP**=120.

The control variables AGE and SEX are considered fixed but do not need to be specified to obtain an odds ratio because the model contains no interaction terms.

The odds ratio is then computed as e to the quantity (SMK* − SMK**) times β_1 plus (PAL* − PAL**) times β_2 plus (SBP* − SBP**) times β_3,

which equals e to (0 − 1) times β_1 plus (25 − 10) times β_2 plus (160 − 120) times β_3,

which equals e to the quantity −1 times β_1 plus 15 times β_2 plus 40 times β_3,

which reduces to e to the quantity $- \beta_1$ plus $15\beta_2$ plus $40\beta_3$.

An estimate of this odds ratio can then be obtained by fitting the model and replacing β_1, β_2, and β_3 by their corresponding estimates β_1 "hat," β_2 "hat," and β_3 "hat." Thus, ROR "hat" equals e to the quantity $- \beta_1$ "hat" plus $15\beta_2$ "hat" plus $40\beta_3$ "hat."

ANOTHER EXAMPLE

$E^* = (SMK^*=1, PAL^*=25, SBP^*=160)$

$E^{**} = (SMK^{**}=1, PAL^{**}=5, SBP^{**}=200)$

controlling for AGE and SEX

$$ROR_{E^* \text{ vs. } E^{**}} = \exp\left[(1-1)\beta_1 + (25-5)\beta_2 + (160-200)\beta_3\right]$$

$$= \exp\left[(0)\beta_1 + (20)\beta_2 + (-40)\beta_3\right]$$

$$= \exp(20\beta_2 - 40\beta_3)$$

As a second example, suppose we compare a smoker who has a PAL score of 25 and a systolic blood pressure of 160 to a smoker who has a PAL score of 5 and a systolic blood pressure of 200, again controlling for AGE and SEX.

The ROR is then computed as e to the quantity $(1 - 1)$ times β_1 plus $(25 - 5)$ times β_2 plus $(160 - 200)$ times β_3, which equals e to 0 times β_1 plus 20 times β_2 plus -40 times β_3, which reduces to e to the quantity $20\beta_2$ minus $40\beta_3$.

VI. The Model and Odds Ratio for Several Exposure Variables with Confounders and Interaction

We now consider a final situation involving **several exposure variables, confounders** (i.e., V's), and **interaction variables** (i.e., W's), where the W's go into the model as product terms with one of the E's.

EXAMPLE: The Variables

$E_1 = SMK, E_2 = PAL, E_3 = SBP$

$V_1 = AGE = W_1, V_2 = SEX = W_2$

$E_1 W_1 = SMK \times AGE, \quad E_1 W_2 = SMK \times SEX$

$E_2 W_1 = PAL \times AGE, \quad E_2 W_2 = PAL \times SEX$

$E_3 W_1 = SBP \times AGE, \quad E_3 W_2 = SBP \times SEX$

As an example, we again consider the three exposures SMK, PAL, and SBP and the two control variables AGE and SEX. We add to this list product terms involving each exposure with each control variable. These product terms are shown here.

EXAMPLE: The Model

$$\text{logit } P(\mathbf{X}) = \alpha + \beta_1 SMK + \beta_2 PAL + \beta_3 SBP$$
$$+ \gamma_1 AGE + \gamma_2 SEX$$
$$+ SMK(\delta_{11}AGE + \delta_{12}SEX)$$
$$+ PAL(\delta_{21}AGE + \delta_{22}SEX)$$
$$+ SBP(\delta_{31}AGE + \delta_{32}SEX)$$

The corresponding model is given by logit $P(\mathbf{X})$ equals α plus β_1 times SMK plus β_2 times PAL plus β_3 times SBP plus the sum of V terms involving AGE and SEX plus SMK times the sum of δ times W terms, where the W's are AGE and SEX, plus PAL times the sum of additional δ times W terms, plus SBP times the sum of additional δ times W terms. Here the δ's are coefficients of interaction terms involving one of the three exposure variables—either SMK, PAL, or SEX—and one of the two control variables—either AGE or SEX.

EXAMPLE: The Odds Ratio

E^* vs. E^{**}

$E^* = (SMK^*=0, PAL^*=25, SBP^*=160)$

$E^{**} = (SMK^{**}=1, PAL^{**}=10, SBP^{**}=120)$

To obtain an odds ratio expression for this model, we again must identify two specifications of the collection of exposure variables to be compared. We have referred to these specifications generally by the bold terms E^* and E^{**}. In the above example, E^* is defined by SMK$^* = 0$, PAL$^* = 25$, and SBP$^* = 160$, whereas E^{**} is defined by SMK$^{**} = 1$, PAL$^{**} = 10$, and SBP$^{**} = 120$.

ROR (no interaction): β's only

ROR (interaction): β's and δ's

The previous odds ratio formula that we gave for several exposures but no interaction involved only β coefficients for the exposure variables. Because the model we are now considering contains interaction terms, the corresponding odds ratio will involve not only the β coefficients, but also δ coefficients for all interaction terms involving one or more exposure variables.

EXAMPLE (continued)

$$ROR_{E^* \text{ vs. } E^{**}} = \exp[(SMK^* - SMK^{**})\beta_1$$
$$+ (PAL^* - PAL^{**})\beta_2$$
$$+ (SBP^* - SBP^{**})\beta_3$$
$$+ \delta_{11}(SMK^* - SMK^{**})AGE$$
$$+ \delta_{12}(SMK^* - SMK^{**})SEX$$
$$+ \delta_{21}(PAL^* - PAL^{**})AGE$$
$$+ \delta_{22}(PAL^* - PAL^{**})SEX$$
$$+ \delta_{31}(SBP^* - SBP^{**})AGE$$

$$ROR = \exp[(0 - 1)\beta_1 + (25 - 10)\beta_2$$
$$+ (160 - 120)\beta_3$$

$\widehat{\text{interaction with SMK}}$

$$+ \delta_{11}(0 - 1)AGE + \delta_{12}(0 - 1)SEX$$

$\widehat{\text{interaction with PAL}}$

$$+ \delta_{21}(25 - 10)AGE + \delta_{22}(25 - 10)SEX$$

$\widehat{\text{interaction with SBP}}$

$$+ \delta_{31}(160 - 120)AGE + \delta_{32}(160 - 120)SEX$$

$$= \exp(-\beta_1 + 15\beta_2 + 40\beta_3$$
$$- \delta_{11}AGE - \delta_{12}SEX$$
$$+ 15\delta_{21}AGE + 15\delta_{22}SEX$$
$$+ 40\delta_{31}AGE + 40\delta_{32}SEX)$$

$$= \exp(-\beta_1 + 15\beta_2 + 40\beta_3$$
$$+ AGE(-\delta_{11} + 15\delta_{21} + 40\delta_{31})$$
$$+ SEX(-\delta_{12} + 15\delta_{22} + 40\delta_{32})]$$

The odds ratio formula for our example then becomes e to the quantity $(SMK^* - SMK^{**})$ times β_1 plus $(PAL^* - PAL^{**})$ times β_2 plus $(SBP^* - SBP^{**})$ times β_3 plus the sum of terms involving a δ coefficient times the difference between E^* and E^{**} values of one of the exposures times a W variable.

For example, the first of the interaction terms is δ_{11} times the difference $(SMK^* - SMK^{**})$ times AGE, and the second of these terms is δ_{12} times the difference $(SMK^* - SMK^{**})$ times SEX.

When we substitute into the odds ratio formula the values for E^* and E^{**}, we obtain the expression e to the quantity $(0 - 1)$ times β_1 plus $(25 - 10)$ times β_2 plus $(160 - 120)$ times β_3 plus several terms involving interaction coefficients denoted as δ's.

The first set of these terms involves interactions of AGE and SEX with SMK. These terms are δ_{11} times the difference $(0 - 1)$ times AGE plus δ_{12} times the difference $(0 - 1)$ times SEX. The next set of δ terms involves interactions of AGE and SEX with PAL. The last set of δ terms involves interactions of AGE and SEX with SBP.

After subtraction, this expression reduces to the expression shown here at the left.

We can simplify this expression further by factoring out AGE and SEX to obtain e to the quantity minus β_1 plus 15 times β_2 plus 40 times β_3 plus AGE times the quantity minus δ_{11} plus 15 times δ_{21} plus 40 times δ_{31} plus SEX times the quantity minus δ_{12} plus 15 times δ_{22} plus 40 times δ_{32}.

EXAMPLE (continued)

Note: Specify AGE and SEX to get a numerical value.

e.g., AGE = 35, SEX = 1:

$$\widehat{ROR} = \exp\left[-\hat{\beta}_1 + 15\hat{\beta}_2 + 40\hat{\beta}_3\right.$$

AGE

SEX

$$+35\left(-\hat{\delta}_{11} + 15\hat{\delta}_{21} + 40\hat{\delta}_{31}\right)$$

$$+1\left(-\hat{\delta}_{12} + 15\hat{\delta}_{22} + 40\hat{\delta}_{32}\right)\Big]$$

$$\widehat{ROR} = \exp\left(-\hat{\beta}_1 + 15\hat{\beta}_2 + 40\hat{\beta}_3\right.$$
$$-35\hat{\delta}_{11} + 525\hat{\delta}_{21} + 1400\hat{\delta}_{31}$$
$$-\hat{\delta}_{12} + 15\hat{\delta}_{22} + 40\hat{\delta}_{32}\Big)$$

General model
   Several exposures
   Confounders
   Effect modifiers

$$\text{logit } P(\mathbf{X}) = \alpha + \beta_1 E_1 + \beta_2 E_2 + \ldots + \beta_q E_q$$

$$+ \sum_{i=1}^{p_1} \gamma_i V_i$$

$$+ E_1 \sum_{j=1}^{p_2} \delta_{1j} W_j + E_2 \sum_{j=1}^{p_2} \delta_{2j} W_j$$

$$+ \ldots + E_q \sum_{j=1}^{p_2} \delta_{qj} W_j$$

Note that this expression tells us that once we have fitted the model to the data to obtain estimates of the β and δ coefficients, we must specify values for the effect modifiers AGE and SEX before we can get a numerical value for the odds ratio. In other words, the odds ratio will give a different numerical value depending on which values we specify for the effect modifiers AGE and SEX.

For instance, if we choose AGE equals 35 and SEX equals 1 say, for females, then the estimated odds ratio becomes the expression shown here.

This odds ratio expression can alternatively be written as e to the quantity minus β_1 "hat" plus 15 times β_2 "hat" plus 40 times β_3 "hat" minus 35 times δ_{11} "hat" plus 525 times δ_{21} "hat" plus 1400 times δ_{31} "hat" minus δ_{12} "hat" plus 15 times δ_{22} "hat" plus 40 times δ_{32} "hat." This expression will give us a single numerical value for 35-year-old females once the model is fitted and estimated coefficients are obtained.

We have just worked through a specific example of the odds ratio formula for a model involving several exposure variables and controlling for both confounders and effect modifiers. To obtain a general odds ratio formula for this situation, we first need to write the model in general form.

This expression is given by the logit of $P(\mathbf{X})$ equals α plus β_1 times E_1 plus β_2 times E_2, and so on up to β_q times E_q plus the usual set of V terms of the form $\gamma_i V_i$ plus the sum of additional terms, each having the form of an exposure variable times the sum of δ times W terms. The first of these interaction expressions is given by E_1 times the sum of δ_{1j} times W_j, where E_1 is the first exposure variable, δ_{1j} is an unknown coefficient, and W_j is the jth effect modifying variable. The last of these terms is E_q times the sum of δ_{qj} times W_j, where E_q is the last exposure variable, δ_{qj} is an unknown coefficient, and W_j is the jth effect modifying variable.

We assume the same W_j for each exposure variable

e.g., AGE and SEX are W's for each E.

Note that this model assumes that the same effect modifying variables are being considered for each exposure variable in the model, as illustrated in our preceding example above with AGE and SEX.

A more general model can be written that allows for different effect modifiers corresponding to different exposure variables, but for convenience, we limit our discussion to a model with the same modifiers for each exposure variable.

Odds ratio for several E's:

$$\mathbf{E}^* = (E_1^*, E_2^*, \ldots, E_q^*)$$
$$\text{vs.}$$
$$\mathbf{E}^{**} = (E_1^{**}, E_2^{**}, \ldots, E_q^{**})$$

To obtain an odds ratio expression for the above model involving several exposures, confounders, and interaction terms, we again must identify two specifications of the exposure variables to be compared. We have referred to these specifications generally by the bold terms $\mathbf{E}^*$ and $\mathbf{E}^{**}$. Group $\mathbf{E}^*$ is specified by the variable values E_1^*, E_2^*, and so on up to E_q^*; group $\mathbf{E}^{**}$ is specified by a different collection of values E_1^{**}, E_2^{**}, and so on up to E_q^{**}.

General Odds Ratio Formula:

$$\text{ROR}_{\mathbf{E}^* \text{ vs. } \mathbf{E}^{**}} = \exp\left[\left(E_1^* - E_1^{**}\right)\beta_1\right.$$
$$+ \left(E_2^* - E_2^{**}\right)\beta_2 + \cdots + \left(E_q^* - E_q^{**}\right)\beta_q$$
$$+ \left(E_1^* - E_1^{**}\right)\sum_{j=1}^{p_2} \delta_j W_j$$
$$+ \left(E_2^* - E_2^{**}\right)\sum_{j=1}^{p_2} \delta_{2j} W_j$$
$$\left. + \cdots + \left(E_q^* - E_q^{**}\right)\sum_{j=1}^{p_2} \delta_{qj} W_j\right]$$

The general odds ratio formula for comparing two such specifications, $\mathbf{E}^*$ versus $\mathbf{E}^{**}$, is given by the formula ROR equals e to the quantity $(E_1^* - E_1^{**})$ times β_1 plus $(E_2^* - E_2^{**})$ times β_2, and so on up to $(E_q^* - E_q^{**})$ times β_q plus the sum of terms of the form $(\mathbf{E}^* - \mathbf{E}^{**})$ times the sum of δ times W, where each of these latter terms correspond to interactions involving a different exposure variable.

EXAMPLE: $q = 3$

$$\begin{aligned} ROR_{E^* \text{ vs. } E^{**}} = \exp\big[&(SMK^* - SMK^{**})\beta_1 \\ &+ (PAL^* - PAL^{**})\beta_2 + (SBP^* - SBP^{**})\beta_3 \\ &+ \delta_{11}(SMK^* - SMK^{**})AGE \\ &+ \delta_{12}(SMK^* - SMK^{**})SEX \\ &+ \delta_{21}(PAL^* - PAL^{**})AGE \\ &+ \delta_{22}(PAL^* - PAL^{**})SEX \\ &+ \delta_{31}(SBP^* - SBP^{**})AGE \\ &+ \delta_{32}(SBP^* - SBP^{**})SEX\big] \end{aligned}$$

- AGE and SEX controlled as V's as well as W's
- ROR's depend on values of W's (AGE and SEX)

In our previous example using this formula, there are q equals three exposure variables (namely, SMK, PAL, and SBP), two confounders (namely, AGE and SEX), which are in the model as V variables, and two effect modifiers (also AGE and SEX), which are in the model as W variables. The odds ratio expression for this example is shown here again.

This odds ratio expression does not contain coefficients for the confounding effects of AGE and SEX. Nevertheless, these effects are being controlled because AGE and SEX are contained in the model as V variables in addition to being W variables.

Note that for this example, as for any model containing interaction terms, the odds ratio expression will yield different values for the odds ratio depending on the values of the effect modifiers—in this case, AGE and SEX—that are specified.

SUMMARY

Chapters up to this point:

1. Introduction
2. Important Special Cases
✓ 3. Computing the Odds Ratio
4. Maximum Likelihood (ML) Techniques: An Overview
5. Statistical Inferences Using ML Techniques

This presentation is now complete. We have described how to compute the odds ratio for an arbitrarily coded single exposure variable that may be dichotomous, ordinal, or interval. We have also described the odds ratio formula when the exposure variable is a polytomous nominal variable like occupational status. And, finally, we have described the odds ratio formula when there are several exposure variables, controlling for confounders without interaction terms and controlling for confounders together with interaction terms.

In the next chapter (Chapter 4), we consider how the method of maximum likelihood is used to estimate the parameters of the logistic model. And in Chapter 5, we describe statistical inferences using ML techniques.

Detailed Outline

I. Overview (pages 76–77)

A. Focus: computing OR for E, D relationship adjusting for confounding and effect modification.

B. Review of the special case—the E, V, W model:

   i. The model: $\text{logit } P(\mathbf{X}) = \alpha + \beta E + \sum_{i=1}^{p_1} \gamma_i V_i + E\sum_{j=1}^{p_2} \delta_j W_j$.

   ii. Odds ratio formula for the E, V, W model, where E is a $(0, 1)$ variable:

   $$\text{ROR}_{E=1 \text{ vs. } E=0} = \exp\left(\beta + \sum_{j=1}^{p_2}\delta_j W_j\right).$$

II. Odds ratio for other codings of a dichotomous E (pages 77–79)

A. For the E, V, W model with E coded as $E = a$ if exposed and as $E = b$ if unexposed, the odds ratio formula becomes

   $$\text{ROR}_{E=1 \text{ vs. } E=0} = \exp\left[(a-b)\beta + (a-b)\sum_{j=1}^{p_2}\delta_j W_j\right]$$

B. Examples: $a = 1$, $b = 0$: $\text{ROR} = \exp(\beta)$
   $a = 1$, $b = -1$: $\text{ROR} = \exp(2\beta)$
   $a = 100$, $b = 0$: $\text{ROR} = \exp(100\beta)$

C. Final computed odds ratio has the same value provided the correct formula is used for the corresponding coding scheme, even though the coefficients change as the coding changes.

D. Numerical example from Evans County study.

III. Odds ratio for arbitrary coding of E (pages 79–82)

A. For the E, V, W model where $\mathbf{E}^*$ and $\mathbf{E}^{**}$ are any two values of E to be compared, the odds ratio formula becomes

   $$\text{ROR}_{E^* \text{ vs. } E^{**}} = \exp\left[(E^*-E^{**})\beta + (E^*-E^{**})\sum_{j=1}^{p_2}\delta_j W_j\right]$$

B. Examples: E = SSU = social support status (0–5)
   E = SBP = systolic blood pressure (interval).

C. No interaction odds ratio formula:
   $\text{ROR}_{E^* \text{ vs. } E^{**}} = \exp[(E^* - E^{**})\beta]$.

D. Interval variables, e.g., SBP: Choose values for comparison that represent clinically meaningful categories, e.g., quintiles.

IV. The model and odds ratio for a nominal exposure variable (no interaction case) (pages 82–84)

A. No interaction model involving a nominal exposure variable with k categories:

$$\text{logit } P(\mathbf{X}) = \alpha + \beta_1 E_1 + \beta_2 E_2 + \cdots + \beta_{k-1} E_{k-1} + \sum_{i=1}^{p_1} \gamma_i V_i$$

where $E_1, E_2, \ldots, E_{k-1}$ denote $k - 1$ dummy variables that distinguish the k categories of the nominal exposure variable denoted as $\mathbf{E}$, i.e.,
$E_i = 1$ if category i or 0 if otherwise.

B. Example of model involving $k = 4$ categories of occupational status:

$$\text{logit } P(\mathbf{X}) = \alpha + \beta_1 \text{OCC}_1 + \beta_2 \text{OCC}_2 + \beta_3 \text{OCC}_3 + \sum_{i=1}^{p_1} \gamma_i V_i$$

where OCC_1, OCC_2, and OCC_3 denote $k - 1 = 3$ dummy variables that distinguish the four categories of occupation.

C. Odds ratio formula for no interaction model involving a nominal exposure variable:

$$\text{ROR}_{\mathbf{E}^* \text{ vs. } \mathbf{E}^{**}} = \exp\left[\begin{array}{c} \left(E_1^* - E_1^{**}\right)\beta_1 + \left(E_2^* - E_2^{**}\right)\beta_2 \\ + \cdots + \left(E_{k-1}^* - E_{k-1}^{**}\right)\beta_{k-1} \end{array} \right]$$

where $\mathbf{E}^* = (E_1^*, E_2^*, \ldots, E_{k-1}^*)$ and $\mathbf{E}^{**} = (E_1^{**}, E_2^{**}, \ldots, E_{k-1}^{**})$ are two specifications of the set of dummy variables for $\mathbf{E}$ to be compared.

D. Example of odds ratio involving $k = 4$ categories of occupational status:

$$\text{ROR}_{\text{OCC}^* \text{ vs. } \text{OCC}^{**}} = \exp\left[\begin{array}{c} \left(\text{OCC}_1^* - \text{OCC}_1^{**}\right)\beta_1 + \left(\text{OCC}_2^* - \text{OCC}_2^{**}\right)\beta_2 \\ + \left(\text{OCC}_3^* - \text{OCC}_3^{**}\right)\beta_3 \end{array} \right].$$

V. The model and odds ratio for several exposure variables (no interaction case) (pages 85–87)

A. The model:

$$\text{logit } P(\mathbf{X}) = \alpha + \beta_1 E_1 + \beta_2 E_2 + \cdots + \beta_q E_q + \sum_{i=1}^{p_1} \gamma_i V_i$$

where $E_1, E_2, \ldots, E_q$ denote q exposure variables of interest.

B. Example of model involving three exposure variables:

$$\text{logit } P(\mathbf{X}) = \alpha + \beta_1 \text{SMK} + \beta_2 \text{PAL} + \beta_3 \text{SBP} + \sum_{i=1}^{p_1} \gamma_i V_i.$$

C.   The odds ratio formula for the general no interaction model:

$$\text{ROR}_{\mathbf{E^*} \text{ vs. } \mathbf{E^{**}}} = \exp\left[\left(E_1^* - E_1^{**}\right)\beta_1 + \left(E_2^* - E_2^{**}\right)\beta_2 + \cdots + \left(E_q^* - E_q^{**}\right)\beta_q\right]$$

where $\mathbf{E^*} = (E_1^*, E_2^*, \ldots, E_q^*)$ and $\mathbf{E^{**}} = (E_1^{**}, E_2^{**}, \ldots, E_q^{**})$ are two specifications of the collection of exposure variables to be compared.

D.   Example of odds ratio involving three exposure variables:

$$\text{ROR}_{\mathbf{E^*} \text{ vs. } \mathbf{E^{**}}} = \exp\left[\left(\text{SMK}^* - \text{SMK}^{**}\right)\beta_1 + \left(\text{PAL}^* - \text{PAL}^{**}\right)\beta_2\right.$$
$$\left. + \left(\text{SBP}^* - \text{SBP}^{**}\right)\beta_3\right].$$

VI.   The model and odds ratio for several exposure variables with confounders and interaction (pages 87–91)

A.   An example of a model with three exposure variables:

$$\text{logit P}(\mathbf{X}) = \alpha + \beta_1 \text{SMK} + \beta_2 \text{PAL} + \beta_3 \text{SBP} + \gamma_1 \text{AGE} + \gamma_2 \text{SEX}$$
$$+ \text{SMK}(\delta_{11}\text{AGE} + \delta_{12}\text{SEX}) + \text{PAL}(\delta_{21}\text{AGE} + \delta_{22}\text{SEX})$$
$$+ \text{SBP}(\delta_{31}\text{AGE} + \delta_{32}\text{SEX}).$$

B.   The odds ratio formula for the above model:

$$\text{ROR}_{\mathbf{E^*} \text{ vs. } \mathbf{E^{**}}} = \exp\left[\left(\text{SMK}^* - \text{SMK}^{**}\right)\beta_1 + \left(\text{PAL}^* - \text{PAL}^{**}\right)\beta_2 + \left(\text{SBP}^* - \text{SBP}^{**}\right)\beta_3\right.$$
$$+ \delta_{11}\left(\text{SMK}^* - \text{SMK}^{**}\right)\text{AGE} + \delta_{12}\left(\text{SMK}^* - \text{SMK}^{**}\right)\text{SEX}$$
$$+ \delta_{21}\left(\text{PAL}^* - \text{PAL}^{**}\right)\text{AGE} + \delta_{22}\left(\text{PAL}^* - \text{PAL}^{**}\right)\text{SEX}$$
$$\left. + \delta_{31}\left(\text{SBP}^* - \text{SBP}^{**}\right)\text{AGE} + \delta_{32}\left(\text{SBP}^* - \text{SBP}^{**}\right)\text{SEX}\right]$$

C.   The general model:

$$\text{logit P}(\mathbf{X}) = \alpha + \beta_1 E_1 + \beta_2 E_2 + \cdots + \beta_q E_q + \sum_{i=1}^{p_1} \gamma_i V_i + E_1 \sum_{j=1}^{p_2} \delta_{1j} W_j$$
$$+ E_2 \sum_{j=1}^{p_2} \delta_{2j} W_j + \cdots + E_q \sum_{j=1}^{p_2} \delta_{qj} W_j$$

D.   The general odds ratio formula:

$$\text{ROR}_{\mathbf{E^*} \text{ vs. } \mathbf{E^{**}}} = \exp\left[\left(E_1^* - E_1^{**}\right)\beta_1 + \left(E_2^* - E_2^{**}\right)\beta_2 + \cdots + \left(E_q^* - E_q^{**}\right)\beta_q\right.$$
$$+ \left(E_1^* - E_1^{**}\right)\sum_{j=1}^{p_2} \delta_{1j} W_j + \left(E_2^* - E_2^{**}\right)\sum_{j=1}^{p_2} \delta_{2j} W_j$$
$$\left. + \cdots + \left(E_q^* - E_q^{**}\right)\sum_{j=1}^{p_2} \delta_{qj} W_j\right]$$

Practice Exercises

Given the model

$$\text{logit } P(\mathbf{X}) = \alpha + \beta E + \gamma_1(\text{SMK}) + \gamma_2(\text{HPT}) + \delta_1(E \times \text{SMK}) + \delta_2(E \times \text{HPT}),$$

where SMK (smoking status) and HPT (hypertension status) are dichotomous variables,

Answer the following true or false questions (circle T or F):

T   F   1.   If E is coded as (0=unexposed, 1=exposed), then the odds ratio for the E, D relationship that controls for SMK and HPT is given by

$$\exp[\beta + \delta_1(E \times \text{SMK}) + \delta_2(E \times \text{HPT})].$$

T   F   2.   If E is coded as $(-1, 1)$, then the odds ratio for the E, D relationship that controls for SMK and HPT is given by
$$\exp[2\beta + 2\delta_1(\text{SMK}) + 2\delta_2(\text{HPT})].$$

T   F   3.   If there is no interaction in the above model and E is coded as $(-1, 1)$, then the odds ratio for the E, D relationship that controls for SMK and HPT is given by $\exp(\beta)$.

T   F   4.   If the correct odds ratio formula for a given coding scheme for E is used, then the estimated odds ratio will be the same regardless of the coding scheme used.

Given the model

$$\text{logit } P(\mathbf{X}) = \alpha + \beta(\text{CHL}) + \gamma(\text{AGE}) + \delta(\text{AGE} \times \text{CHL}),$$

where CHL and AGE are continuous variables,

Answer the following true or false questions (circle T or F):

T   F   5.   The odds ratio that compares a person with CHL=200 to a person with CHL=140 controlling for AGE is given by $\exp(60\beta)$.

T   F   6.   If we assume no interaction in the above model, the expression $\exp(\beta)$ gives the odds ratio for describing the effect of one unit change in CHL value, controlling for AGE.

Suppose a study is undertaken to compare the lung cancer risks for samples from three regions (urban, suburban and rural) in a certain state, controlling for the potential confounding and effect modifying effects of AGE, smoking status (SMK), RACE, and SEX.

7.   State the logit form of a logistic model that treats region as a polytomous exposure variable and controls for the confounding effects of AGE, SMK, RACE, and SEX. (Assume no interaction involving any covariates with exposure.)

8. For the model of Exercise 7, give an expression for the odds ratio for the E, D relationship that compares urban with rural persons, controlling for the four covariates.

9. Revise your model of Exercise 7 to allow effect modification of each covariate with the exposure variable. State the logit form of this revised model.

10. For the model of Exercise 9, give an expression for the odds ratio for the E, D relationship that compares urban with rural persons, controlling for the confounding and effect modifying effects of the four covariates.

11. Given the model

$$\text{logit } P(\mathbf{X}) = \alpha + \beta_1(\text{SMK}) + \beta_2(\text{ASB}) + \gamma_1(\text{AGE})$$
$$+ \delta_1(\text{SMK} \times \text{AGE}) + \delta_2(\text{ASB} \times \text{AGE}),$$

where SMK is a $(0, 1)$ variable for smoking status, ASB is a $(0, 1)$ variable for asbestos exposure status, and AGE is treated continuously,

Circle the (one) correct choice among the following statements:

a. The odds ratio that compares a smoker exposed to asbestos to a non-smoker not exposed to asbestos, controlling for age, is given by $\exp(\beta_1 + \beta_2 + \delta_1 + \delta_2)$.

b. The odds ratio that compares a nonsmoker exposed to asbestos to a nonsmoker unexposed to asbestos, controlling for age, is given by $\exp[\beta_2 + \delta_2(\text{AGE})]$.

c. The odds ratio that compares a smoker exposed to asbestos to a smoker unexposed to asbestos, controlling for age, is given by $\exp[\beta_1 + \delta_1(\text{AGE})]$.

d. The odds ratio that compares a smoker exposed to asbestos to a non-smoker exposed to asbestos, controlling for age, is given by $\exp[\beta_1 + \delta_1(\text{AGE}) + \delta_2(\text{AGE})]$.

e. None of the above statements is correct.

Test

1. Given the following logistic model
    $$\text{logit } P(\mathbf{X}) = \alpha + \beta\text{CAT} + \gamma_1\text{AGE} + \gamma_2\text{CHL},$$
    where CAT is a dichotomous exposure variable and AGE and CHL are continuous, **answer the following questions concerning the odds ratio that compares exposed to unexposed persons controlling for the effects of AGE and CHL:**

    a. Give an expression for the odds ratio for the E, D relationship, assuming that CAT is coded as (0=low CAT, 1=high CAT).

    b. Give an expression for the odds ratio, assuming CAT is coded as (0, 5).

    c. Give an expression for the odds ratio, assuming that CAT is coded as $(-1, 1)$.

    d. Assuming that the same data set is used for computing odds ratios described in parts a–c above, what is the relationship among odds ratios computed by using the three different coding schemes of parts a–c?

    e. Assuming the same data set as in part d above, what is the relationship between the β's that are computed from the three different coding schemes?

2. Suppose the model in Question 1 is revised as follows:
    $$\text{logit } P(\mathbf{X}) = \alpha + \beta\text{CAT} + \gamma_1\text{AGE} + \gamma_2\text{CHL} + \text{CAT}(\delta_1\text{AGE} + \delta_2\text{CHL}).$$

    For this revised model, answer the same questions as given in parts a–e of Question 1.

    a.

    b.

    c.

    d.

    e.

3. Given the model
    $$\text{logit } P(\mathbf{X}) = \alpha + \beta\text{SSU} + \gamma_1\text{AGE} + \gamma_2\text{SEX} + \text{SSU}(\delta_1\text{AGE} + \delta_2\text{SEX}),$$
    where SSU denotes "social support score" and is an ordinal variable ranging from 0 to 50, **answer the following questions about the above model:**

    a. Give an expression for the odds ratio that compares a person who has SSU=50 to a person who has SSU=0, controlling for AGE and SEX.

b. Give an expression for the odds ratio that compares a person who has SSU=10 to a person who has SSU=0, controlling for AGE and SEX.

c. Give an expression for the odds ratio that compares a person who has SSU=20 to a person who has SSU=10, controlling for AGE and SEX.

d. Assuming that the same data set is used for parts b and c, what is the relationship between the odds ratios computed in parts b and c?

4. Suppose the variable SSU in Question 3 is partitioned into three categories denoted as *low, medium,* and *high.*

a. Revise the model of Question 3 to give the logit form of a logistic model that treats SSU as a nominal variable with three categories (*assume no interaction*).

b. Using your model of part a, give an expression for the odds ratio that compares high to low SSU persons, controlling for AGE and SEX.

c. Revise your model of part a to allow for effect modification of SSU with AGE and with SEX.

d. Revise your odds ratio of part b to correspond to your model of part c.

5. Given the following model

$$\text{logit } P(\mathbf{X}) = \alpha + \beta_1 NS + \beta_2 OC + \beta_3 AFS + \gamma_1 AGE + \gamma_2 RACE,$$

where NS denotes number of sex partners in one's lifetime, OC denotes oral contraceptive use (yes/no), and AFS denotes age at first sexual intercourse experience, **answer the following questions about the above model:**

a. Give an expression for the odds ratio that compares a person who has NS=5, OC=1, and AFS=26 to a person who has NS=5, OC=1, and AFS=16, controlling for AGE and RACE.

b. Give an expression for the odds ratio that compares a person who has NS=200, OC=1, and AFS=26 to a person who has NS=5, OC=1, and AFS=16, controlling for AGE and RACE.

6. Suppose the model in Question 5 is revised to contain interaction terms:

$$\begin{aligned} \text{logit } P(\mathbf{X}) = &\, \alpha + \beta_1 NS + \beta_2 OC + \beta_3 AFS + \gamma_1 AGE + \gamma_2 RACE \\ &+ \delta_{11}(NS \times AGE) + \delta_{12}(NS \times RACE) + \delta_{21}(OC \times AGE) \\ &+ \delta_{22}(OC \times RACE) + \delta_{31}(AFS \times AGE) + \delta_{32}(AFS \times RACE). \end{aligned}$$

For this revised model, answer the same questions as given in parts a and b of Question 5.

a.

b.

Answers to Practice Exercises

1. F: the correct odds ratio expression is $\exp[\beta + \delta_1(SMK) + \delta_2(HPT)]$

2. T

3. F: the correct odds ratio expression is $\exp(2\beta)$

4. T

5. F: the correct odds ratio expression is $\exp[60\beta + 60\delta(AGE)]$

6. T

7. $\text{logit } P(\mathbf{X}) = \alpha + \beta_1 R_1 + \beta_2 R_2 + \gamma_1 AGE + \gamma_2 SMK + \gamma_3 RACE + \gamma_4 SEX$, where R_1 and R_2 are dummy variables indicating region, e.g., $R_1 = (1$ if urban, 0 if other) and $R_2 = (1$ if suburban, 0 if other).

8. When the above coding for the two dummy variables is used, the odds ratio that compares urban with rural persons is given by $\exp(\beta_1)$.

9. $\text{logit } P(\mathbf{X}) = \alpha + \beta_1 R_1 + \beta_2 R_2 + \gamma_1 AGE + \gamma_2 SMK + \gamma_3 RACE + \gamma_4 SEX + R_1(\delta_{11}AGE + \delta_{12}SMK + \delta_{13}RACE + \delta_{14}SEX) + R_2(\delta_{21}AGE + \delta_{22}SMK + \delta_{23}RACE + \delta_{24}SEX).$

10. Using the coding of the answer to Question 7, the revised odds ratio expression that compares urban with rural persons is $\exp(\beta_1 + \delta_{11}AGE + \delta_{12}SMK + \delta_{13}RACE + \delta_{14}SEX).$

11. The correct answer is b.

4

Maximum Likelihood Techniques: An Overview

Introduction

In this chapter, we describe the general maximum likelihood (ML) procedure, including a discussion of likelihood functions and how they are maximized. We also distinguish between two alternative ML methods, called the unconditional and the conditional approaches, and we give guidelines regarding how the applied user can choose between these methods. Finally, we provide a brief overview of how to make statistical inferences using ML estimates.

Abbreviated Outline

The outline below gives the user a preview of the material to be covered by the presentation. Together with the objectives, this outline offers the user an overview of the content of this module. A detailed outline for review purposes follows the presentation.

Objectives   Upon completing this chapter, the learner should be able to:

1. State or recognize when to use unconditional versus conditional ML methods.
2. State or recognize what is a likelihood function.
3. State or recognize that the likelihood functions for unconditional versus conditional ML methods are different.
4. State or recognize that unconditional versus conditional ML methods require different computer programs.
5. State or recognize how an ML procedure works to obtain ML estimates of unknown parameters in a logistic model.
6. Given a logistic model, state or describe two alternative procedures for testing hypotheses about parameters in the model. In particular, describe each procedure in terms of the information used (log likelihood statistic or Z statistic) and the distribution of the test statistic under the null hypothesis (chi square or Z).
7. State, recognize, or describe three types of information required for carrying out statistical inferences involving the logistic model: the value of the maximized likelihood, the variance–covariance matrix, and a listing of the estimated coefficients and their standard errors.
8. Given a logistic model, state or recognize how interval estimates are obtained for parameters of interest; in particular, state that interval estimates are large sample formulae that make use of variance and covariances in the variance–covariance matrix.
9. Given a printout of ML estimates for a logistic model, use the printout information to describe characteristics of the fitted model. In particular, given such a printout, compute an estimated odds ratio for an exposure–disease relationship of interest.

Presentation

I. Overview

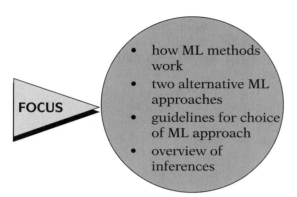

This presentation gives an overview of maximum likelihood (ML) methods as used in logistic regression analysis. We focus on how ML methods work, we distinguish between two alternative ML approaches, and we give guidelines regarding which approach to choose. We also give a brief overview on making statistical inferences using ML techniques.

II. Background about Maximum Likelihood Procedure

Maximum likelihood (ML) estimation

Least squares (LS) estimation: used in classical linear regression

- ML = LS when normality is assumed

Maximum likelihood (ML) estimation is one of several alternative approaches that statisticians have developed for estimating the parameters in a mathematical model. Another well-known and popular approach is **least squares (LS) estimation,** which is described in most introductory statistics courses as a method for estimating the parameters in a classical straight line or multiple linear regression model. ML estimation and least squares estimation are different approaches that happen to give the same results for classical linear regression analyses when the dependent variable is assumed to be normally distributed.

ML estimation

- computer programs available
- general applicability
- used for nonlinear models, e.g., the logistic model

For many years, ML estimation was not widely used because no computer software programs were available to carry out the complex calculations required. However, ML programs have been widely available for the past 10 or so years. Moreover, when compared to least squares, the ML method can be applied in the estimation of complex nonlinear as well as linear models. In particular, because the logistic model is a nonlinear model, ML estimation is the preferred estimation method for logistic regression.

Discriminant function analysis

- previously used for logistic model
- restrictive normality assumptions
- gives biased results—odds ratio too high

Until the recent availability of computer software for ML estimation, the method used to estimate the parameters of a logistic model was **discriminant function analysis.** This method has been shown by statisticians to be essentially a least squares approach. Restrictive normality assumptions on the independent variables in the model are required to make statistical inferences about the model parameters. In particular, if any of the independent variables are dichotomous or categorical in nature, then the discriminant function method tends to give biased results, usually giving estimated odds ratios that are too high.

ML estimation

- no restrictions on independent variables
- preferred to discriminant analysis

ML estimation, on the other hand, requires no restrictions of any kind on the characteristics of the independent variables. Thus, when using ML estimation, the independent variables can be nominal, ordinal, and/or interval. Consequently, ML estimation is to be preferred over discriminant function analysis for fitting the logistic model.

III. Unconditional Versus Conditional Methods

look at unconditional

Two alternative ML approaches

(1) unconditional method
(2) conditional method
- require different computer programs
- user must choose appropriate program

There are actually two alternative ML approaches that can be used to estimate the parameters in a logistic model. These are called the **unconditional method** and the **conditional method.** These two methods require different computer programs. Thus, researchers using logistic regression modeling must decide which of these two programs is appropriate for their data.

Computer Programs

SAS (LOGIST)
BMDP
GLIM
SPSS
EGRET
SPIDA
S+
} unconditional

Until recently, the most widely available computer packages for ML estimation of the logistic model, namely, **SAS's LOGIST** procedure and a corresponding procedure from **BMDP,** have been unconditional procedures. However, now several computer packages are available. A list of packages for unconditional estimation is provided here. At the top of the next page, we list computer packages for conditional estimation.

$$\left.\begin{array}{l} \text{SAS (PECAN)} \\ \text{SAS (PHREG)} \\ \text{EGRET} \\ \text{SPIDA} \\ \text{S+} \end{array}\right\} \text{conditional}$$

Note that there are some newly developed computer packages that include ML estimation techniques. These include **EGRET,** from the Statistics and Epidemiology Research Corporation, Seattle, Washington; **SPIDA,** from Macquarie University in Sydney, Australia; and **S+** from Statistical Sciences, Inc., Seattle, Washington.

The Choice

Unconditional—preferred if number of parameters is *small* relative to number of subjects

Conditional—preferred if number of parameters is *large* relative to number of subjects

In making the choice between **unconditional** and **conditional ML approaches,** the researcher needs to consider the number of parameters in one's model relative to the total number of subjects under study. In general, **unconditional** ML estimation is preferred if the number of parameters in the model is **small** relative to the number of subjects. In contrast, **conditional** ML estimation is preferred if the number of parameters in the model is **large** relative to the number of subjects.

Small vs. large? debatable

Guidelines provided here

Exactly what is small versus what is large is debatable and has not yet nor may ever be precisely determined by statisticians. Nevertheless, we can provide some guidelines for choosing the estimation method.

EXAMPLE: Unconditional Preferred

Cohort study: 10-year follow-up
$n = 700$
D = CHD outcome
E = exposure variable

C_1, C_2, C_3, C_4, C_5 = covariables

$E \times C_1, E \times C_2, E \times C_3, E \times C_4, E \times C_5$ = interaction terms

Number of parameters = 12 (including intercept)

small relative to n = 700

An example of a situation suitable for an unconditional ML program is a large cohort study that does not involve matching, for instance, a study of 700 subjects who are followed for 10 years to determine coronary heart disease status, denoted here as CHD. Suppose, for the analysis of data from such a study, a logistic model is considered involving an exposure variable E, five covariables C_1 through C_5 treated as confounders in the model, and five interaction terms of the form $E \times C_i$, where C_i is the ith covariable.

This model contains a total of 12 parameters, one for each of the variables plus one for the intercept term. Because the number of parameters here is 12 and the number of subjects is 700, this is a situation suitable for using **unconditional ML estimation;** that is, the number of parameters is **small** relative to the number of subjects.

EXAMPLE: Conditional Preferred

Case-control study
  100 matched pairs
  D = lung cancer

Matching variables:
  age, race, sex, location

Other variables:
  SMK (a confounder)
  E (dietary characteristic)

Case-control study
  100 matched pairs

Logistic model for matching:
- uses dummy variables for matching strata
- 99 dummy variables for 100 strata
- E, SMK, and $E \times$ SMK also in model

Number of parameters =

$$\underset{\substack{\uparrow \\ \text{intercept}}}{1} \quad + \quad \underset{\substack{\uparrow \\ \text{dummy} \\ \text{variables}}}{99} \quad + \quad \underset{\substack{\uparrow \\ E, \text{SMK } E \times \text{SMK}}}{3} \quad = \quad \boxed{103}$$

large relative to 100 matched
pairs $\Rightarrow$ $\boxed{n = 200}$

In contrast, consider a case-control study involving 100 matched pairs. Suppose that the outcome variable is lung cancer and that controls are matched to cases on age, race, sex, and location. Suppose also that smoking status, a potential confounder denoted as SMK, is not matched but is nevertheless determined for both cases and controls, and that the primary exposure variable of interest, labeled as E, is some dietary characteristic, such as whether or not a subject has a high-fiber diet.

Because the study design involves matching, a logistic model to analyze this data must control for the matching by using dummy variables to reflect the different matching strata, each of which involves a different matched pair. Assuming the model has an intercept, the model will need 99 dummy variables to incorporate the 100 matched pairs. Besides these variables, the model contains the exposure variable E, the covariable SMK, and perhaps even an interaction term of the form $E \times$ SMK.

To obtain the number of parameters in the model, we must count the one intercept, the coefficients of the 99 dummy variables, the coefficient of E, the coefficient of SMK, and the coefficient of the product term E $\times$ SMK. The total number of parameters is 103. Because there are 100 matched pairs in the study, the total number of subjects is, therefore, 200. This situation requires **conditional ML estimation** because the number of parameters, 103, is quite **large** relative to the number of subjects, 200.

REFERENCE
Chapter 8: Analysis of Matched Data Using Logistic Regression

A detailed discussion of logistic regression for matched data is provided in Chapter 8.

Guidelines

- use *conditional* if matching

- use *unconditional* if no matching and number of variables not too large

The above examples indicate the following guidelines regarding the choice between unconditional and conditional ML methods or programs:

- Use **conditional ML estimation** whenever matching has been done; this is because the model will invariably be large due to the number of dummy variables required to reflect the matching strata.
- Use **unconditional ML estimation** if matching has not been done, provided the total number of variables in the model is not unduly large relative to the number of subjects.

EXAMPLE

Unconditional questionable if
- 10 to 15 confounders
- 10 to 15 product terms

Loosely speaking, this means that if the total number of confounders and the total number of interaction terms in the model are large, say 10 to 15 confounders and 10 to 15 product terms, the number of parameters may be getting too large for the unconditional approach to give accurate answers.

Safe rule:
Use *conditional* when in doubt
- gives unbiased results always
- unconditional may be biased (may overestimate odds ratios)

A safe rule is to use conditional ML estimation whenever in doubt about which method to use, because, theoretically, the conditional approach has been shown by statisticians to give unbiased results always. In contrast, the unconditional approach, when unsuitable, can give biased results and, in particular, can overestimate odds ratios of interest.

Note: *Conditional* is expensive when unconditional is appropriate

Note that use of the conditional approach can get expensive in terms of the cost of computer running time. Consequently, the researcher may save time and money using an unconditional program when the model being used is suitable for the unconditional approach. However, recent advances in computer technology have resulted in much faster computer time for carrying out the conditional approach.

EXAMPLE: Conditional Required

Pair-matched case-control study
measure of effect: OR

As a simple example of the need to use conditional ML estimation for matched data, consider again a pair-matched case-control study such as described above. For such a study design, the measure of effect of interest is an odds ratio for the exposure–disease relationship that adjusts for the variables being controlled.

EXAMPLE: (Continued)

Assume: only variables controlled are matched

Then
$$\widehat{OR}_U = (\widehat{OR}_C)^2$$
$$\underset{\text{biased}}{\uparrow} \quad \underset{\text{correct}}{\uparrow}$$

e.g.,
$$\widehat{OR}_C = 3 \Rightarrow \widehat{OR}_U = (3)^2 = 9$$

If the **only** variables being controlled are those involved in the matching, then the estimate of the odds ratio obtained by using unconditional ML estimation, which we denote by $\widehat{OR}_U$, is the square of the estimate obtained by using conditional ML estimation, which we denote by $\widehat{OR}_C$. Statisticians have shown that the correct estimate of this OR is given by the conditional method, whereas a biased estimate is given by the unconditional method.

Thus, for example, if the conditional ML estimate yields an estimated odds ratio of 3, then the unconditional ML method will yield a very large overestimate of 3 squared, or 9.

R-to-1 matching
$$\Downarrow$$
unconditional is overestimate of (correct) conditional estimate

More generally, whenever matching is used, even R-to-1 matching, where R is greater than 1, the unconditional estimate of the odds ratio that adjusts for covariables will give an overestimate, though not necessarily the square, of the conditional estimate.

Having now distinguished between the two alternative ML procedures, we are ready to describe the ML procedure in more detail and to give a brief overview of how statistical inferences are made using ML techniques.

IV. The Likelihood Function and Its Use in the ML Procedure

$L = L(\underset{\sim}{\theta}) =$ likelihood function

$\underset{\sim}{\theta} = (\theta_1, \theta_2, \ldots, \theta_q)$

To describe the ML procedure, we introduce the likelihood function, L. This is a function of the unknown parameters in one's model and, thus can alternatively be denoted as $L(\underset{\sim}{\theta})$, where $\underset{\sim}{\theta}$ denotes the collection of unknown parameters being estimated in the model. In matrix terminology, the collection $\underset{\sim}{\theta}$ is referred to as a **vector**; its components are the individual parameters being estimated in the model, denoted here as θ_1, θ_2, up through θ_q, where q is the number of individual components.

E, V, W model:

$$\text{logit } P(\mathbf{X}) = \alpha + \beta E + \sum_{i=1}^{p_1} \gamma_i V_i + E \sum_{j=1}^{p_2} \delta_j W_j$$

$$\underset{\sim}{\theta} = (\alpha, \beta, \gamma_1, \gamma_2, \ldots, \delta_1, \delta_2, \ldots)$$

For example, using the E, V, W logistic model previously described and shown here again, the unknown parameters are α, β, the γ_i's, and the δ_j's. Thus, the vector of parameters $\underset{\sim}{\theta}$ has α, β, the γ_i's, and the δ_j's as its components.

$L = L(\theta)$
   = joint probability of observing the data

The **likelihood function** L or $L(\theta)$ **represents the joint probability or likelihood of observing the data that have been collected.** The term "joint probability" means a probability that combines the contributions of all the subjects in the study.

EXAMPLE

$n = 100$ trials
p = probability of success
$x = 75$ successes
$n - x = 25$ failures

Pr (75 successes out of 100 trials) has binomial distribution

$$Pr\ (X = 75 \mid n = 100, p)$$
$$\uparrow$$
$$given$$

$$Pr\ (X = 75 \mid n = 100, p)$$
$$= c \times p^{75} \times (1 - p)^{100 - 75}$$
$$= L(p)$$

As a simple example, in a study involving 100 trials of a new drug, suppose the parameter of interest is the probability of a successful trial, which is denoted by p. Suppose also that, out of the n equal to 100 trials studied, there are x equal to 75 successful trials and $n - x$ equal to 25 failures. The probability of observing 75 successes out of 100 trials is a joint probability and can be described by the binomial distribution. That is, the model is a binomial-based model, which is different from and much less complex than the logistic model.

The binomial probability expression is shown here. This is stated as the probability that X, the number of successes, equals 75 given that there are n equal to 100 trials and that the probability of success on a single trial is p. Note that the vertical line within the probability expression means "given."

This probability is numerically equal to a constant c times p to the 75th power times $1 - p$ to the $100 - 75$ or 25th power. This expression is the likelihood function for this example. It gives the probability of observing the results of the study as a function of the unknown parameters, in this case the single parameter p.

ML method maximizes the likelihood function $L(\theta)$

$$\hat{\underset{\sim}{\theta}} = \left(\hat{\theta}_1, \hat{\theta}_2, \ldots, \hat{\theta}_q\right) = \text{ML estimator}$$

Once the likelihood function has been determined for a given set of study data, **the method of maximum likelihood chooses that estimator of the set of unknown parameters $\underset{\sim}{\theta}$ which maximizes the likelihood function $L(\theta)$.** The estimator is denoted as $\underset{\sim}{\theta}$ "hat" and its components are θ_1 "hat," θ_2 "hat," and so on up through θ_q "hat."

EXAMPLE (Binomial)

ML solution:
$\hat{p}$ maximizes
$L(p) = c \times p^{75} \times (1 - p)^{25}$

In the binomial example described above, the maximum likelihood solution gives that value of the parameter p which maximizes the likelihood expression c times p to the 75th power times $1 - p$ to the 25th power. The estimated parameter here is denoted as p "hat."

EXAMPLE (continued)

Maximum value obtained by solving

$$\frac{dL}{dp} = 0$$

for p:

$\hat{p} = 0.75$ "most likely"

maximum
↓
$$p > \hat{p} = 0.75 \Rightarrow L(p) < L(p = 0.75)$$

e.g.,

binomial formula
$$p = 1 \Rightarrow L(1) = c \times 1^{75} \times (1-1)^{25}$$
$$= 0 < L(0.75)$$

$$\hat{p} = 0.75 = \frac{75}{100}, \text{ a sample proportion}$$

Binomial model

$$\Rightarrow \hat{p} = \frac{X}{n} \text{ is ML estimator}$$

More complicated models ⇒ complex calculations

The standard approach for maximizing an expression like the likelihood function for the binomial example here is to use calculus by setting the derivative dL/dp equal to 0 and solving for the unknown parameter or parameters.

For the binomial example, when the derivative dL/dp is set equal to 0, the ML solution obtained is p "hat" equal to 0.75. Thus, the value 0.75 is the "most likely" value for p in the sense that it maximizes the likelihood function L.

If we substitute into the expression for L a value for p exceeding 0.75, this will yield a smaller value for L than obtained when substituting p equal to 0.75. This is why 0.75 is called the ML estimator. For example, when p equals 1, the value for L using the binomial formula is 0, which is as small as L can get and is, therefore, less than the value of L when p equals the ML value of 0.75.

Note that for the binomial example, the ML value p "hat" equal to 0.75 is simply the sample proportion of the 100 trials which are successful. In other words, for a binomial model, **the sample proportion** always turns out to be the ML estimator of the parameter p. So for this model, it is not necessary to work through the calculus to derive this estimate. However, for models more complicated than the binomial, for example, the logistic model, calculus computations involving derivatives are required and are quite complex.

Maximizing $L(\underset{\sim}{\theta})$ is equivalent to maximizing $\ln L(\underset{\sim}{\theta})$

Solve: $\dfrac{\partial \ln L(\underset{\sim}{\theta})}{\partial \theta_j} = 0, j = 1, 2, \ldots, q$

In general, maximizing the likelihood function $L(\underset{\sim}{\theta})$ is equivalent to maximizing the natural log of $L(\underset{\sim}{\theta})$, which is computationally easier. The components of $\underset{\sim}{\theta}$ are then found as solutions of equations of partial derivatives as shown here. Each equation is stated as the partial derivative of the log of the likelihood function with respect to θ_j equals 0, where θ_j is the jth individual parameter.

q equations in q unknowns require *iterative* solution by computer

If there are q parameters in total, then the above set of equations is a set of q equations in q unknowns. These equations must then be solved iteratively, which is no problem with the right computer program.

Two alternatives:
  unconditional program (L_U)
           vs.
  conditional program (L_C)
  $\uparrow$
  $\nwarrow$ $|$

           likelihoods

Formula for L is built into
  computer programs

User inputs data and
  computer does calculations

L formulae are different for unconditional
  and conditional methods

The unconditional formula:
  (a joint probability)

cases   noncases
$\downarrow$ $\downarrow$

$$L_U = \prod_{l=1}^{m_1} P(\mathbf{X}_l) \prod_{l=m_1+1}^{n} \left[1 - P(\mathbf{X}_l)\right]$$

$P(\mathbf{X})$ = logistic model

$$= \frac{1}{1 + e^{-(\alpha + \Sigma \beta_i X_i)}}$$

$$L_U = \frac{\displaystyle\prod_{l=1}^{n} \exp\left(\alpha + \sum_{i=1}^{k} \beta_i X_{il}\right)}{\displaystyle\prod_{l=1}^{n} \left[1 + \exp\left(\alpha + \sum_{i=1}^{k} \beta_i X_{il}\right)\right]}$$

As described earlier, if the model is logistic, there are **two alternative types** of computer programs to choose from, an **unconditional** versus a **conditional** program. These programs use different likelihood functions, namely, L_U for the unconditional method and L_C for the conditional method.

The formulae for the likelihood functions for both the unconditional and conditional ML approaches are quite complex mathematically. The applied user of logistic regression, however, never has to see the formulae for L in practice because they are built into their respective computer programs. All the user has to do is learn how to input the data and to state the form of the logistic model being fit. Then the program does the heavy calculations of forming the likelihood function internally and maximizing this function to obtain the ML solutions.

Although we do not want to emphasize the particular likelihood formulae for the unconditional versus conditional methods, we do want to describe how these formulae are different. Thus, we briefly show these formulae for this purpose.

The **unconditional formula** is given first, and directly describes the joint probability of the study data as the **product of the joint probability for the cases** (diseased persons) **and the joint probability for the noncases** (nondiseased persons). These two products are indicated by the large Π signs in the formula. We can use these products here by assuming that we have independent observations on all subjects. The probability of obtaining the data for the lth case is given by $P(\mathbf{X}_l)$, where $P(\mathbf{X})$ is the logistic model formula for individual $\mathbf{X}$. The probability of the data for the lth noncase is given by $1 - P(\mathbf{X}_l)$.

When the logistic model formula involving the parameters is substituted into the likelihood expression above, the formula shown here is obtained after a certain amount of algebra is done. Note that this expression for the likelihood function L is a function of the unknown parameters α and the β_i.

The conditional formula:

$$L_C = \frac{\Pr(\text{observed data})}{\Pr(\text{all possible configurations})}$$

The conditional likelihood formula (L_C) reflects the probability of the observed data configuration relative to the probability of all possible configurations of the given data. To understand this, we describe the observed data configuration as a collection of m_1 cases and $n-m_1$ noncases,. We denote the cases by the **X** vectors $\mathbf{X}_1$, $\mathbf{X}_2$, and so on through $\mathbf{X}_{m_1}$ and the noncases by $\mathbf{X}_{m_1+1}$, $\mathbf{X}_{m_1+2}$, through $\mathbf{X}_n$.

m_1 cases: $(\mathbf{X}_1, \mathbf{X}_2, ..., \mathbf{X}_{m_1})$

$n-m_1$ noncases: $(\mathbf{X}_{m_1+1}, \mathbf{X}_{m_1+2}, ..., \mathbf{X}_n)$

$L_C = \Pr(\text{first } m_1 \ \mathbf{X}\text{'s are cases} \mid \text{all possible configurations of } \mathbf{X}\text{'s})$

The above configuration assumes that we have rearranged the observed data so that the m_1 cases are listed first and are then followed in listing by the $n-m_1$ noncases. Using this configuration, the conditional likelihood function gives the probability that the first m_1 of the observations actually go with the cases, given all possible configurations of the above n observations into a set of m_1 cases and a set of $n-m_1$ noncases.

EXAMPLE: Configurations

(1) Last m_1 **X**'s are cases
$$(\mathbf{X}_1, \mathbf{X}_2, ..., \mathbf{X}_n)$$
$\underline{\qquad}$ cases

(2) Cases of **X**'s are in middle of listing
$$(\mathbf{X}_1, \mathbf{X}_2, ..., \mathbf{X}_n)$$
$\underline{\qquad}$ cases

The term **configuration** here refers to one of the possible ways that the observed set of **X** vectors can be partitioned into m_1 cases and $n-m_1$ noncases. In example 1 here, for instance, the last m_1 **X** vectors are the cases and the remaining **X**'s are noncases. In example 2, however, the m_1 cases are in the middle of the listing of all **X** vectors.

Possible configurations
= combinations of n things taken m_1 at a time
= $C^n_{m_1}$

The number of possible configurations is given by the number of combinations of n things taken m_1 at a time, which is denoted mathematically by the expression shown here, where the C in the expression denotes combinations.

$$L_C = \frac{\prod_{l=1}^{m_1} P(\mathbf{X}_l) \prod_{l=m_1+1}^{n} \left[1 - P(\mathbf{X}_l)\right]}{\sum_u \left\{ \prod_{l=1}^{m_1} P(\mathbf{X}_{ul}) \prod_{l=m_1+1}^{n} \left[1 - P(\mathbf{X}_{ul})\right] \right\}}$$

versus

$$L_U = \prod_{l=1}^{m_1} P(\mathbf{X}_l) \prod_{l=m_1+1}^{n} \left[1 - P(\mathbf{X}_l)\right]$$

The formula for the conditional likelihood is then given by the expression shown here. The numerator is exactly the same as the likelihood for the unconditional method. The denominator is what makes the conditional likelihood different from the unconditional likelihood. Basically, the denominator sums the joint probabilities for all possible configurations of the m observations into m_1 cases and $n-m_1$ noncases. Each configuration is indicated by the u in the L_C formula.

$$L_C = \frac{\prod_{l=1}^{m_1} \exp\left(\sum_{i=1}^{k} \beta_i X_{li}\right)}{\sum_u \left[\prod_{l=1}^{m_1} \exp\left(\sum_{i=1}^{k} \beta_i X_{lui}\right)\right]}$$

Note: α drops out of L_C

Conditional program:

- estimate β's
- does not estimate α (nuisance parameter)

Note: OR involves only β's

Case-control study: cannot estimate α

$$L_U \neq L_C$$

direct joint probability   does not require estimating nuisance parameters

Stratified data, e.g., matching,
$\Downarrow$
many nuisance parameters

100 nuisance parameters

are not estimated using L_C unnecessarily estimated using L_U

When the logistic model formula involving the parameters is substituted into the conditional likelihood expression above, the resulting formula shown here is obtained. This formula is not the same as the unconditional formula shown earlier. Moreover, in the conditional formula, the intercept parameter α has dropped out of the likelihood.

The removal of the intercept α from the conditional likelihood is important because it means that when a conditional ML program is used, estimates are obtained only for the β_i coefficients in the model and not for α. Because the usual focus of a logistic regression analysis is to estimate an odds ratio, which involves the β's and not α, we usually do not care about estimating α and, therefore, consider α to be a nuisance parameter.

In particular, if the data come from a case-control study, we cannot estimate α because we cannot estimate risk, and the conditional likelihood function does not allow us to obtain any such estimate.

Regarding likelihood functions, then, we have shown that the unconditional and conditional likelihood functions involve different formulae. The unconditional formula has the theoretical advantage in that it is developed directly as a joint probability of the observed data. The conditional formula has the advantage that it does not require estimating nuisance parameters like α.

If the data are stratified, as, for example, by matching, it can be shown that there are as many nuisance parameters as there are matched strata. Thus, for example, if there are 100 matched pairs, then 100 nuisance parameters do not have to be estimated when using conditional estimation, whereas these 100 parameters would be unnecessarily estimated when using unconditional estimation.

Matching:

Unconditional $\Rightarrow$ biased estimates of β's
Conditional $\Rightarrow$ unbiased estimates of β's

If we consider the other parameters in the model for matched data, that is, the β's, the unconditional likelihood approach gives biased estimates of the β's, whereas the conditional approach gives unbiased estimates of the β's.

V. Overview on Statistical Inferences for Logistic Regression

Chapter 5: Statistical Inferences Using Maximum Likelihood Techniques

Statistical inferences involve
- testing hypotheses
- obtaining confidence intervals

We have completed our description of the ML method in general, distinguished between unconditional and conditional approaches, and distinguished between their corresponding likelihood functions. We now provide a brief overview of how statistical inferences are carried out for the logistic model. A detailed discussion of statistical inferences is given in the next chapter.

Once the ML estimates have been obtained, the next step is to use these estimates to make **statistical inferences** concerning the exposure–disease relationships under study. This step includes testing hypotheses and obtaining confidence intervals for parameters in the model.

Quantities required from computer output:

Inference-making can be accomplished through the use of two quantities that are part of the output provided by standard ML estimation programs.

The first of these quantities is the **maximized likelihood value,** which is simply the numerical value of the likelihood function L when the ML estimates (θ "hat") are substituted for their corresponding parameter values (θ). This value is called $L(\theta$ "hat") in our earlier notation.

(1) Maximized likelihood value $L(\hat{\theta})$
(2) Estimated variance–covariance matrix

covariances off the diagonal

$\hat{V}(\hat{\theta}) =$
variances on diagonal

The second quantity is the **estimated variance–covariance matrix.** This matrix, V "hat" of θ "hat," has as its diagonal the estimated variances of each of the ML estimates. The values off the diagonal are the covariances of pairs of ML estimates. The reader may recall that the covariance between two estimates is the correlation times the standard error of each estimate.

Note: $\widehat{\text{cov}}(\hat{\theta}_1, \hat{\theta}_2) = r_{12}s_1 s_2$

namely
maximum
and θ = vector
of params
we use.

Importance of $\hat{V}(\hat{\underset{\sim}{\theta}})$:

   inferences require accounting for
   variability and covariability

(3) Variable listing

Variable	ML Coefficient	S. E.
Intercept	$\hat{\alpha}$	$s_{\hat{\alpha}}$
X_1	$\hat{\beta}_1$	$s_{\hat{\beta}_1}$
.	.	
.	.	
.	.	
X_k	$\hat{\beta}_k$	$s_{\hat{\beta}_k}$

The variance–covariance matrix is important because the information contained in it is used in the computations required for hypothesis testing and confidence interval estimation.

In addition to the maximized likelihood value and the variance–covariance matrix, other information is also provided as part of the output. This information typically includes, as shown here, **a listing of each variable followed by its ML estimate and standard error.** This information provides another way to carry out hypothesis testing and interval estimation. Moreover, this listing gives the primary information used for calculating odds ratio estimates and predicted risks. The latter can only be done, however, if the study has a follow-up design.

An example of ML computer output giving the above information is provided here. This output considers study data on a cohort of 609 white males in Evans County, Georgia, who were followed for 9 years to determine coronary heart disease (CHD) status. The output considers a logistic model involving eight variables, which are denoted as CAT (catecholamine level), AGE, CHL (cholesterol level), ECG (electrocardiogram abnormality status), SMK (smoking status), HPT (hypertension status), CC, and CH. The latter two variables are product terms of the form CC=CAT × CHL and CH=CAT × HPT.

The exposure variable of interest here is the variable CAT, and the five covariables of interest, that is, the C's are AGE, CHL, ECG, SMK, and HPT. Using our E, V, W model framework described in the review section, we have E equals CAT, the five covariables equal to the V's, and two W variables, namely, CHL and HPT.

The output information includes -2 times the natural log of the maximized likelihood value, which is 347.78, and a listing of each variable followed by its ML estimate and standard error. We will show the variance–covariance matrix shortly.

EXAMPLE

Cohort study—Evans County, GA

$$n = 609 \text{ white males}$$
$$9\text{-year follow-up}$$
$$D = \text{CHD status}$$

Output: $-2 \ln \hat{L} = 347.28$

Variable	ML Coefficient	S. E.
Intercept	−4.0474	1.2549
CAT	−12.6809	3.1042
AGE	0.0349	0.0161
CHL	−0.0055	0.0042
ECG	0.3665	0.3278
SMK	0.7735	0.3272
HPT	1.0468	0.3316
CC	−2.3299	0.7422
CH	0.0691	0.0143

V's { AGE, CHL, ECG, SMK, HPT }

CC = CAT × CHL and CH = CAT × HPT
           W's

EXAMPLE (continued)

$\widehat{OR}$ considers coefficients of CAT, CC, and CH

$\widehat{OR} = \exp(\hat{\beta} + \hat{\delta}_1 CHL + \hat{\delta}_2 HPT)$

where

$\hat{\beta} = -12.6809$
$\hat{\delta}_1 = 0.0691$
$\hat{\delta}_2 = -2.3299$

$\widehat{OR} = \exp[-12.6809 + 0.0691 CHL + (-2.3299)HPT]$

Must specify:

CHL and HPT

effect modifiers

Note: $\widehat{OR}$ different for different values specified for CHL and HPT

$\widehat{OR} = \exp(-12.6809 + 0.0691 CHL - 2.3299 HPT)$

		HPT	
		0	1
	200	3.12	0.30
CHL	220	12.44	1.21
	240	49.56	4.82

CHL = 200, HPT = 0: $\widehat{OR} = 3.12$

CHL = 220, HPT = 1: $\widehat{OR} = 1.21$

$\widehat{OR}$ adjusts for AGE, CHL, ECG, SMK, and HPT (the V variables)

We now consider how to use the information provided to obtain an estimated odds ratio for the fitted model. Because this model contains the product terms CC equal to CAT × CHL, and CH equal to CAT × HPT, the estimated odds ratio for the effect of CAT must consider the coefficients of these terms as well as the coefficient of CAT.

The formula for this estimated odds ratio is given by the exponential of the quantity β "hat" plus δ_1 "hat" times CHL plus δ_2 "hat" times HPT, where β "hat" equals -12.6809 is the coefficient of CAT, δ_1 "hat" equals 0.0691 is the coefficient of the interaction term CC and δ_2 "hat" equals -2.3299 is the coefficient of the interaction term CH.

Plugging the estimated coefficients into the odds ratio formula yields the expression: e to the quantity -12.6809 plus 0.0691 times CHL plus -2.3299 times HPT.

To obtain a numerical value from this expression, it is necessary to specify a value for CHL and a value for HPT. Different values for CHL and HPT will, therefore, yield different odds ratio values, as should be expected because the model contains interaction terms.

The table shown here illustrates different odds ratio estimates that can result from specifying different values of the effect modifiers. In this table, the values of CHL are 200, 220, and 240; the values of HPT are 0 and 1, where 1 denotes a person who has hypertension. The cells within the table give the estimated odds ratios computed from the above expression for the odds ratio for different combinations of CHL and HPT.

For example, when CHL equals 200 and HPT equals 0, the estimated odds ratio is given by 3.12; when CHL equals 220 and HPT equals 1, the estimated odds ratio is 1.21. Note that each of the estimated odds ratios in this table describes the association between CAT and CHD adjusted for the five covariables AGE, CHL, ECG, SMK, and HPT because each of the covariables is contained in the model as V variables.

$\widehat{OR}$'s = point estimators

Variability of $\widehat{OR}$ considered for statistical inferences

The estimated model coefficients and the corresponding odds ratio estimates that we have just described are point estimates of unknown population parameters. Such point estimates have a certain amount of variability associated with them, as illustrated, for example, by the standard errors of each estimated coefficient provided in the output listing. We consider the variability of our estimates when we make statistical inferences about parameters of interest.

Two types of inferences:
(1)   testing hypotheses
(2)   interval estimation

We can use two kinds of inference-making procedures. One is testing hypotheses about certain parameters; the other is deriving interval estimates of certain parameters.

EXAMPLES

(1)   Test for H_0:   OR = 1

(2)   Test for significant interaction, e.g., $\delta_1 \neq 0$?

(3)   Interval estimate: 95% confidence interval for $OR_{CAT, CHD}$ controlling for 5 V's and 2 W's

Interaction: must specify W's e.g., 95% confidence interval when CAT = 220 and HPT = 1

As an example of a test, we may wish to test the null hypothesis that an odds ratio is equal to the null value.

Or, as another example, we may wish to test for evidence of significant interaction, for instance, whether one or more of the coefficients of the product terms in the model are significantly nonzero.

As an example of an interval estimate, we may wish to obtain a 95% confidence interval for the adjusted odds ratio for the effect of CAT on CHD, controlling for the five V variables and the two W variables. Because this model contains interaction terms, we need to specify the values of the W's to obtain numerical values for the confidence limits. For instance, we may want the 95% confidence interval when CHL equals 220 and HPT equals 1.

Two testing procedures:
(1)   *Likelihood ratio test*: a chi-square statistic using $-2 \ln \hat{L}$.
(2)   *Wald test*: a Z test using standard errors listed with each variable.

When using ML estimation, we can carry out hypothesis testing by using one of two procedures, the **likelihood ratio test** and the **Wald test.** The likelihood ratio test is a chi-square test which makes use of maximized likelihood values such as shown in the output. The Wald test is a Z test; that is, the test statistic is approximately standard normal. The Wald test makes use of the standard errors shown in the listing of variables and associated output information. Each of these procedures is described in detail in the next chapter.

Large samples: both procedures give approximately the same results

Small or moderate samples: different results possible; likelihood ratio test preferred

Confidence intervals
- use large sample formulae
- use variance–covariance matrix

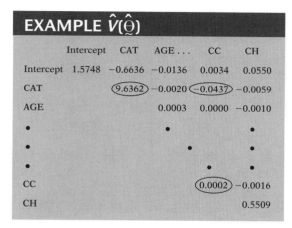

EXAMPLE $\hat{V}(\hat{\theta})$

	Intercept	CAT	AGE ...	CC	CH
Intercept	1.5748	−0.6636	−0.0136	0.0034	0.0550
CAT		9.6362	−0.0020	−0.0437	−0.0059
AGE			0.0003	0.0000	−0.0010
•		•			•
•			•		•
•				•	•
CC				0.0002	−0.0016
CH					0.5509

No interaction: variance only

Interaction: variances and covariances

Both testing procedures should give approximately the **same answer in large samples but may give different results in small or moderate samples.** In the latter case, statisticians prefer the likelihood ratio test to the Wald test.

Confidence intervals are carried out by using large sample formulae that make use of the information in the variance–covariance matrix, which includes the variances of estimated coefficients together with the covariances of pairs of estimated coefficients.

An example of the estimated variance–covariance matrix is given here. Note, for example, that the variance of the coefficient of the CAT variable is 9.6362, the variance for the CC variable is 0.0002, and the covariance of the coefficients of CAT and CC is −0.0437.

If the model being fit contains no interaction terms and if the exposure variable is a (0, 1) variable, then only a variance estimate is required for computing a confidence interval. If the model contains interaction terms, then both variance and covariance estimates are required; in this latter case, the computations required are much more complex than when there is no interaction.

SUMMARY

Chapters up to this point:
1. Introduction
2. Important Special Cases
3. Computing the Odds Ratio
✓ 4. ML Techniques: An Overview

This presentation is now complete. In summary, we have described how ML estimation works, have distinguished between unconditional and conditional methods and their corresponding likelihood functions, and have given an overview of how to make statistical inferences using ML estimates.

We suggest that the reader review the material covered here by reading the summary outline that follows. Then you may work the practice exercises and test.

5. Statistical Inferences Using ML Techniques

In the next chapter, we give a detailed description of how to carry out both testing hypotheses and confidence interval estimation for the logistic model.

**Detailed
Outline**

I. **Overview** (page 104)

Focus
- how ML methods work
- two alternative ML approaches
- guidelines for choice of ML approach
- overview of statistical inferences

II. **Background about maximum likelihood procedure**
(pages 104–105)
   A. Alternative approaches to estimation: least squares (LS), maximum likelihood (ML), and discriminant function analysis.
   B. ML is now the preferred method—computer programs now available; general applicability of ML method to many different types of models.

III. **Unconditional versus conditional methods** (pages 105–109)
   A. Require different computer programs; user must choose appropriate program.
   B. **Unconditional** preferred if number of parameters **small** relative to number of subjects, whereas **conditional** preferred if number of parameters **large** relative to number of subjects.
   C. Guidelines: use conditional if matching; use unconditional if no matching and number of variables not too large; when in doubt, use conditional—always unbiased.

IV. **The likelihood function and its use in the ML procedure**
(pages 109–115)
   A. $L = L(\theta)$ = likelihood function; gives joint probability of observing the data as a function of the set of unknown parameters given by $\underset{\sim}{\theta} = (\theta_1, \theta_2, \ldots, \theta_q)$.
   B. ML method maximizes the likelihood function $L(\underset{\sim}{\theta})$.
   C. ML solutions solve a system of q equations in q unknowns; this system requires an *iterative* solution by computer.
   D. Two alternative likelihood functions for logistic regression: unconditional (L_U) and conditional (L_C); formulae are built into unconditional and conditional programs.
   E. User inputs data and computer does calculations.
   F. Conditional likelihood reflects the probability of observed data configuration relative to the probability of all possible configurations of the data.
   G. Conditional program estimates β's but not α (nuisance parameter).
   H. Matched data: unconditional gives biased estimates, whereas conditional gives unbiased estimates.

V. Overview on statistical inferences for logistic regression
(pages 115–119)
  A. Two types of inferences: testing hypotheses and confidence interval estimation.
  B. Three items obtained from computer output for inferences:
    i. Maximized likelihood value $L(\hat{\theta})$;
    ii. Estimated variance–covariance matrix $\hat{V}(\hat{\theta})$: variances on diagonal and covariances on the off-diagonal;
    iii. Variable listing with ML estimates and standard errors.
  C. Two testing procedures:
    i. **Likelihood ratio test:** a chi-square statistic using $-2 \ln \hat{L}$.
    ii. **Wald test:** a Z test using standard errors listed with each variable.
  D. Both testing procedures give approximately same results with large samples; with small samples, different results are possible; likelihood ratio test is preferred.
  E. Confidence intervals: use large sample formulae that involve variances and covariances from variance–covariance matrix.

Practice Exercises

True or False (Circle T or F)

T F   1. When estimating the parameters of the logistic model, least squares estimation is the preferred method of estimation.

T F   2. Two alternative maximum likelihood approaches are called unconditional and conditional methods of estimation.

T F   3. The conditional approach is preferred if the number of parameters in one's model is small relative to the number of subjects in one's data set.

T F   4. Conditional ML estimation should be used to estimate logistic model parameters if matching has been carried out in one's study.

T F   5. Unconditional ML estimation gives unbiased results always.

T F   6. The likelihood function $L(\theta)$ represents the joint probability of observing the data that has been collected for analysis.

T F   7. The maximum likelihood method maximizes the function $\ln L(\theta)$.

T F   8. The likelihood function formulae for both the unconditional and conditional approaches are the same.

T F   9. The maximized likelihood value $L(\hat{\theta})$ is used for confidence interval estimation of parameters in the logistic model.

T F 10. The likelihood ratio test is the preferred method for testing hypotheses about parameters in the logistic model.

Test

True or False (Circle T or F)

T F   1. Maximum likelihood estimation is preferred to least squares estimation for estimating the parameters of the logistic and other nonlinear models.

T F   2. If discriminant function analysis is used to estimate logistic model parameters, biased estimates can be obtained that result in estimated odds ratios that are too high.

T F   3. In a case-control study involving 1200 subjects, a logistic model involving 1 exposure variable, 3 potential confounders, and 3 potential effect modifiers is to be estimated. Assuming no matching has been done, the preferred method of estimation for this model is conditional ML estimation.

T F   4. Until recently, the most widely available computer packages for fitting the logistic model have used unconditional procedures.

T F   5. In a matched case-control study involving 50 cases and 2-to-1 matching, a logistic model used to analyze the data will contain a small number of parameters relative to the total number of subjects studied.

T F   6. If a likelihood function for a logistic model contains 10 parameters, then the ML solution solves a system of 10 equations in 10 unknowns by using an iterative procedure.

T F   7. The conditional likelihood function reflects the probability of the observed data configuration relative to the probability of all possible configurations of the data.

T F   8. The nuisance parameter α is not estimated using an unconditional ML program.

T F   9. The likelihood ratio test is a chi-square test that uses the maximized likelihood value $\hat{L}$ in its computation.

T F   10. The Wald test and the likelihood ratio test of the same hypothesis give approximately the same results in large samples.

T F   11. The variance–covariance matrix printed out for a fitted logistic model gives the variances of each variable in the model and the covariances of each pair of variables in the model.

T F   12. Confidence intervals for odds ratio estimates obtained from the fit of a logistic model use large sample formulae that involve variances and possibly covariances from the variance–covariance matrix.

The printout given below comes from a matched case-control study of 313 women in Sydney, Australia (Brock et al., *J. Nat. Cancer Inst.*, 1988), to assess the etiologic role of sexual behaviors and dietary factors on the development of cervical cancer. Matching was done on age and socioeconomic status. The outcome variable is cervical cancer status (yes/no), and the independent variables considered here (all coded as 1, 0) are vitamin C intake (VITC, high/low), the number of lifetime sexual partners (NSEX, high/low), age at first intercourse (SEXAGE, old/young), oral contraceptive pill use (PILLM ever/never), and smoking status (CSMOK, ever/never).

Variable	Coefficient	S.E.	OR	*P*	95% Confidence Interval	
VITC	−0.24411	0.14254	0.7834	.086	0.5924	1.0359
NSEX	0.71902	0.16848	2.0524	.000	1.4752	2.8555
SEXAGE	−0.19914	0.25203	0.8194	.426	0.5017	1.3383
PILLM	0.39447	0.19004	1.4836	.037	1.0222	2.1532
CSMOK	1.59663	0.36180	4.9364	.000	2.4290	10.0318

MAX LOG LIKELIHOOD = −73.5088

Using the above printout, answer the following questions:

13. What method of estimation should have been used to fit the logistic model for this data set? Explain.

14. Why don't the variables age and socioeconomic status appear in the printout?

15. Describe how to compute the odds ratio for the effect of pill use in terms of an estimated regression coefficient in the model. Interpret the meaning of this odds ratio.

16. What odds ratio is described by the value *e* to −0.24411? Interpret this odds ratio.

17. State two alternative ways to describe the null hypothesis appropriate for testing whether the odds ratio described in Question 16 is significant.

18. What is the 95% confidence interval for the odds ratio described in Question 16, and what parameter is being estimated by this interval?

19. The *P*-values given in the table correspond to Wald test statistics for each variable adjusted for the others in the model. The appropriate *Z* statistic is computed by dividing the estimated coefficient by its standard error. What is the *Z* statistic corresponding to the *P*-value of .086 for the variable VITC?

20. For what purpose is the quantity denoted as MAX LOG LIKELIHOOD used?

Answers to Practice Exercises

1. F: ML estimation is preferred
2. T
3. F: conditional is preferred if number of parameters is large
4. T
5. F: conditional gives unbiased results
6. T
7. T
8. F: L_U and L_C are different
9. F: The variance–covariance matrix is used for confidence interval estimation
10. T

5

Statistical Inferences Using Maximum Likelihood Techniques

Introduction

We begin our discussion of statistical inference by describing the computer information required for making inferences about the logistic model. We then introduce examples of three logistic models that we use to describe hypothesis testing and confidence interval estimation procedures. We consider models with no interaction terms first, and then we consider how to modify procedures when there is interaction. Two types of testing procedures are given, namely, the likelihood ratio test and the Wald test. Confidence interval formulae are provided that are based on large sample normality assumptions. A final review of all inference procedures is described by way of a numerical example.

Abbreviated Outline

The outline below gives the user a preview of the material to be covered by the presentation. A detailed outline for review purposes follows the presentation.

Objectives   Upon completion of this chapter, the learner should be able to:

1.   State the **null hypothesis** for testing the significance of a collection of one or more variables in terms of regression coefficients of a given logistic model.

2.   Describe how to carry out a **likelihood ratio test** for the significance of one or more variables in a given logistic model.

3.   Use computer information for a fitted logistic model to carry out a likelihood ratio test for the significance of one or more variables in the model.

4.   Describe how to carry out a **Wald test** for the significance of a single variable in a given logistic model.

5.   Use computer information for a fitted logistic model to carry out a Wald test for the significance of a single variable in the model.

6.   Describe how to compute a **95% confidence interval** for an odds ratio parameter that can be estimated from a given logistic model when

     a. the model contains no interaction terms;

     b. the model contains interaction terms.

7.   Use computer information for a fitted logistic model to compute a 95% confidence interval for an odds ratio expression estimated from the model when

     a. the model contains no interaction terms;

     b. the model contains interaction terms.

Presentation

I. Overview

Previous chapter:

- how ML methods work
- unconditional vs. conditional approaches

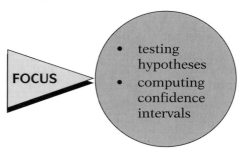

FOCUS

- testing hypotheses
- computing confidence intervals

In the previous chapter, we described how ML methods work in general and we distinguished between two alternative approaches to estimation—the unconditional and the conditional approach.

In this chapter, we describe how statistical inferences are made using ML techniques in logistic regression analyses. We focus on procedures for testing hypotheses and computing confidence intervals about logistic model parameters and odds ratios derived from such parameters.

II. Information for Making Statistical Inferences

Quantities required from output:

(1)   Maximized likelihood value:
$L(\hat{\theta})$

(2)   Estimated variance–covariance matrix:
$\hat{V}(\hat{\theta})$

covariances off the diagonal

$\hat{V}(\hat{\theta}) =$
variances on diagonal

Once ML estimates have been obtained, these estimates can be used to make statistical inferences concerning the exposure–disease relationships under study. Three quantities are required from the output provided by standard ML estimation programs.

The first of these quantities is the **maximized likelihood value,** which is the numerical value of the likelihood function L when the ML estimates are substituted for their corresponding parameter values; this value is called L of θ "hat" in our earlier notation.

The second quantity is the **estimated variance–covariance matrix,** which we denote as V "hat" of θ "hat."

The estimated variance-covariance matrix has on its diagonal the estimated variances of each of the ML estimates. The values off the diagonal are the covariances of pairs of ML estimates.

$$\widehat{\text{cov}}\left(\hat{\theta}_1,\hat{\theta}_2\right) = r_{12}s_1s_2$$

The reader may recall that the covariance between two estimates is the correlation times the standard errors of each estimate.

Importance of $\hat{V}(\hat{\theta})$:

inferences require variances and covariances

The variance–covariance matrix is important because hypothesis testing and confidence interval estimation require variances and sometimes covariances for computation.   — we will be using wall~

(3) Variable listing:

Variable	ML Coefficient	S.E.
Intercept	$\hat{\alpha}$	$s_{\hat{\alpha}}$
X_1	$\hat{\beta}_1$	$s_{\hat{\beta}_1}$
•	•	•
•	•	•
•	•	•
X_k	$\hat{\beta}_k$	$s_{\hat{\beta}_k}$

In addition to the maximized likelihood value and the variance–covariance matrix, other information is also provided as part of the output. This typically includes, as shown here, **a listing of each variable followed by its ML estimate and standard error.** This information provides another way of carrying out hypothesis testing and confidence interval estimation, as we will describe shortly. Moreover, this listing gives the primary information used for calculating odds ratio estimates and predicted risks. The latter can only be done, however, provided the study has a follow-up type of design.

III. Models for Inference-making

Model 1: $\text{logit } P_1(\mathbf{X}) = \alpha + \beta_1 X_1 + \beta_2 X_2$

Model 2: $\text{logit } P_2(\mathbf{X}) = \alpha + \beta_1 X_1 + \beta_2 X_2 + \beta_3 X_3$

Model 3: $\text{logit } P_3(\mathbf{X}) = \alpha + \beta_1 X_1 + \beta_2 X_2 + \beta_3 X_3$
$$+ \beta_4 X_1 X_3 + \beta_5 X_2 X_3$$

$\hat{L}_1, \hat{L}_2, \hat{L}_3$ are $\hat{L}$'s for models 1–3
   ML Values

$\hat{L}_1 \le \hat{L}_2 \le \hat{L}_3$

To illustrate how statistical inferences are made using the above information, we consider the following three models, each written in logit form. Model 1 involves two variables X_1 and X_2. Model 2 contains these same two variables and a third variable X_3. Model 3 contains the same three X's as in model 2 plus two additional variables, which are the product terms $X_1 X_3$ and $X_2 X_3$.

Let L_1 "hat," L_2 "hat," and L_3 "hat" denote the maximized likelihood values based on fitting models 1, 2, and 3, respectively. Note that the fitting may be done either by unconditional or conditional methods, depending on which method is more appropriate for the model and data set being considered.

Because the more parameters a model has, the better it fits the data, it follows that L_1 "hat" must be less than or equal to L_2 "hat," which, in turn, must be less than or equal to L_3 "hat."

After all the model could spit out $\beta_3 = 0$ say, and so M2 will always be better than M1 (has more scope). You are trying to maximise a function with additional variables ... more scope.

$\hat{L}$ similar to R^2

This relationship among the L "hat"s is similar to the property in classical multiple linear regression analyses that the more parameters a model has, the higher is the R square statistic for the model. In other words, the maximized likelihood value L "hat" is similar to R square, in that the higher the L "hat," the better the fit.

$\ln \hat{L}_1 \leq \ln \hat{L}_2 \leq \ln \hat{L}_3$

It follows from algebra that if L_1 "hat" is less than or equal to L_2 "hat," which is less than L_3 "hat," then the same inequality relationship holds for the natural logarithms of these L "hat"s.

$-2 \ln \hat{L}_3 \leq -2 \ln \hat{L}_2 \leq -2 \ln \hat{L}_1$

However, if we multiply each log of L "hat" by -2, then the inequalities switch around so that $-2 \ln L_3$ "hat" is less than or equal to $-2 \ln L_2$ "hat," which is less than $-2 \ln L_1$ "hat."

$-2 \ln \hat{L} =$ log likelihood statistic

used in likelihood ratio (LR) test

The statistic $-2 \ln L_1$ "hat" is called the **log likelihood statistic** for model 1, and similarly, the other two statistics are the log likelihood statistics for their respective models. These statistics are important because they can be used to test hypotheses about parameters in the model using what is called a **likelihood ratio test,** which we now describe.

IV. The Likelihood Ratio Test

$-2 \ln L_1 - \left(-2 \ln L_2 \right) = LR$

is approximate chi square

df = difference in number of parameters (degrees of freedom)

Statisticians have shown that the difference between log likelihood statistics for two models, one of which is a special case of the other, has an approximate chi-square distribution in large samples. Such a test statistic is called a **likelihood ratio** or **LR** statistic. The degrees of freedom (df) for this chi-square test are equal to the difference between the number of parameters in the two models.

Model 1: $\mathrm{logit}\, P_1(\mathbf{X}) = \alpha + \beta_1 X_1 + \beta_2 X_2$

Model 2: $\mathrm{logit}\, P_2(\mathbf{X}) = \alpha + \beta_1 X_1 + \beta_2 X_2 + \beta_3 X_3$

Note: special case = subset

model 1 special case of model 2
model 2 special case of model 3

Note that one model is considered a special case of another if one model contains a subset of the parameters in the other model. For example, model 1 above is a special case of model 2; also, model 2 is a special case of model 3.

LR statistic (like F statistic) compares two models:
full model = larger model
reduced model = smaller model

In general, the likelihood ratio statistic, like an F statistic in classical multiple linear regression, requires the identification of two models to be compared, one of which is a special case of the other. The larger model is sometimes called the **full model** and the smaller model is sometimes called the **reduced model;** that is, the reduced model is obtained by setting certain parameters in the full model equal to zero.

H_0: parameters in full model equal to zero

df = number of parameters set equal to zero

The set of parameters in the full model that is set equal to zero specify the null hypothesis being tested. Correspondingly, the degrees of freedom for the likelihood ratio test are equal to the number of parameters in the larger model that must be set equal to zero to obtain the smaller model.

EXAMPLE

Model 1 vs. Model 2

Model 2 (full model):

$\quad \text{logit } P_2(\mathbf{X}) = \alpha + \beta_1 X_1 + \beta_2 X_2 + \beta_3 X_3$

Model 1 (reduced model):

$\quad \text{logit } P_1(\mathbf{X}) = \alpha + \beta_1 X_1 + \beta_2 X_2$

$H_0: \quad \beta_3 = 0$ (similar to partial F)

Model 2:

$\quad \text{logit } P_2(\mathbf{X}) = \alpha + \beta_1 X_1 + \beta_2 X_2 + \beta_3 X_3$

Suppose $X_3 = E(0, 1)$ and X_1, X_2 confounders.

Then $OR = e^{\beta_3}$

$H_0: \beta_3 = 0 \Leftrightarrow H_0: OR = e^0 = 1$

$LR = -2 \ln \hat{L}_1 - \left(-2 \ln \hat{L}_2\right)$

As an example of a likelihood ratio test, let us now compare model 1 with model 2. Because model 2 is the larger model, we can refer to model 2 as the full model and to model 1 as the reduced model. The additional parameter in the full model that is not part of the reduced model is β_3, the coefficient of the variable X_3. Thus, the null hypothesis that compares models 1 and 2 is stated as β_3 equal to 0. This is similar to the null hypothesis for a partial F test in classical multiple linear regression analysis.

Now consider model 2, and suppose that the variable X_3 is a $(0, 1)$ exposure variable E and that the variables X_1 and X_2 are confounders. Then the odds ratio for the exposure–disease relationship that adjusts for the confounders is given by e to β_3.

Thus, in this case, testing the null hypothesis that β_3 equals 0 is equivalent to testing the null hypothesis that the adjusted odds ratio for the effect of exposure is equal to e to 0, or 1.

To test this null hypothesis, the corresponding likelihood ratio statistic is given by the difference $-2 \ln L_1$ "hat" minus $-2 \ln L_2$ "hat."

EXAMPLE (continued)

ratio of likelihoods

$$-2 \ln \hat{L}_1 - \left(-2 \ln \hat{L}_2\right) = -2 \ln\left(\frac{\hat{L}_1}{\hat{L}_2}\right)$$

LR approximate χ^2 variable with df = 1 if n large

Algebraically, this difference can also be written as -2 times the natural log of the ratio of L_1 "hat" divided by L_2 "hat," shown on the right-hand side of the equation here. This latter version of the test statistic is a ratio of maximized likelihood values; this explains why the test is called the likelihood ratio test.

The likelihood ratio statistic for this example has approximately a chi-square distribution if the study size is large. The degrees of freedom for the test is one because, when comparing models 1 and 2, only one parameter, namely, β_3, is being set equal to zero under the null hypothesis.

How the LR test works:

If X_3 makes a large contribution, then $\hat{L}_2$ much greater than $\hat{L}_1$

We now describe how the likelihood ratio test works and why the test statistic is approximately chi square. We consider what the value of the test statistic would be if the additional variable X_3 makes an extremely large contribution to the risk of disease over that already contributed by X_1 and X_2. Then, it follows that the maximized likelihood value $\hat{L}_2$ "hat" is much larger than the maximized likelihood value L_1 "hat."

If $\hat{L}_2$ much larger than $\hat{L}_1$, then

$$\frac{\hat{L}_1}{\hat{L}_2} \approx 0$$

If L_2 "hat" is much larger than L_1 "hat," then the ratio L_1 "hat" divided by L_2 "hat" becomes a very small fraction; that is, this ratio approaches 0.

[**Note** : $\ln_e$ (fraction) = negative]

$$\Rightarrow \ln\left(\frac{\hat{L}_1}{\hat{L}_2}\right) \approx \ln(0) = -\infty$$

Now the natural log of any fraction is a negative number. As this fraction approaches 0, the log of the fraction, which is negative, approaches the log of 0, which is $-\infty$.

$$\Rightarrow LR = -2 \ln\left(\frac{\hat{L}_1}{\hat{L}_2}\right) \approx \infty$$

Thus, X_3 highly significant $\Rightarrow$ LR large and positive

If we multiply the log likelihood ratio by -2, we then get a number that approaches $+\infty$. Thus, the likelihood ratio statistic for a highly significant X_3 variable is large and positive and approaches $+\infty$. This is exactly the type of result expected for a chi-square statistic.

If X_3 makes no contribution, then

$$\hat{L}_2 \approx \hat{L}_1$$

$$\Rightarrow \frac{\hat{L}_1}{\hat{L}_2} \approx 1$$

$$\Rightarrow \text{LR} \approx -2\ln(1) = -2 \times 0 = 0$$

Thus, X_3 nonsignificant $\Rightarrow$ LR approximate 0

$$0 \leq \text{LR} \leq \infty$$
$$\uparrow \qquad \uparrow$$
N.S.   S.

Similar to chi square (χ^2)

LR approximate χ^2 if n large

How large? no precise answer

In contrast, consider the value of the test statistic if the additional variable makes no contribution whatsoever to the risk of disease over and above that contributed by X_1 and X_2. This would mean that the maximized likelihood value L_2 "hat" is essentially equal to the maximized likelihood value L_1 "hat."

Correspondingly, the ratio L_1 "hat" divided by L_2 "hat" is approximately equal to 1. Therefore, the likelihood ratio statistic is approximately equal to -2 times the natural log of 1, which is 0, because the log of 1 is 0. Thus, the likelihood ratio statistic for a highly non-significant X_3 variable is approximately 0. This, again, is what one would expect from a chi-square statistic.

In summary, the likelihood ratio statistic, regardless of which two models are being compared, yields a value that lies between 0, when there is extreme nonsignificance, and $+\infty$, when there is extreme significance. This is the way a chi-square statistic works.

Statisticians have shown that the likelihood ratio statistic can be considered approximately chi square, provided that the number of subjects in the study is large. How large is large, however, has never been precisely documented, so the applied researcher has to have as large a study as possible and/or hope that the number of study subjects is large enough.

As another example of a likelihood ratio test, we consider a comparison of model 2 with model 3. Because model 3 is larger than model 2, we now refer to model 3 as the full model and to model 2 as the reduced model.

EXAMPLE

Model 2: $\text{logit } P_2(\mathbf{X}) = \alpha + \beta_1 X_1 + \beta_2 X_2 + \beta_3 X_3$

(reduced model)

Model 3: $\text{logit } P_3(\mathbf{X}) = \alpha + \beta_1 X_1 + \beta_2 X_2 + \beta_3 X_3$

(full model) $\qquad\qquad + \beta_4 X_1 X_3 + \beta_5 X_2 X_3$

EXAMPLE (continued)

H_0: $\beta_4 = \beta_5 = 0$

(similar to multiple-partial F test)

H_A: β_4 and/or β_5 are not zero

$X_3 = E$

X_1, X_2 confounders

$X_1 X_3, X_2 X_3$ interaction terms

H_0: $\beta_4 = \beta_5 = 0 \Leftrightarrow H_0$: no interaction with E

$$LR = -2 \ln \hat{L}_2 - \left(-2 \ln \hat{L}_3\right) = -2 \ln \left(\frac{\hat{L}_2}{\hat{L}_3}\right)$$

which is approx. χ^2 with two df under

H_0: $\beta_4 = \beta_5 = 0$

$-2 \ln \hat{L}_2$, $\; -2 \ln \hat{L}_3$

$\uparrow$ $\uparrow$

computer prints these
separately

There are two additional parameters in the full model that are not part of the reduced model; these are β_4 and β_5, the coefficients of the product variables $X_1 X_3$ and $X_2 X_3$, respectively. Thus, the null hypothesis that compares models 2 and 3 is stated as β_4 equals β_5 equals 0. This is similar to the null hypothesis for a multiple-partial F test in classical multiple linear regression analysis. The alternative hypothesis here is that β_4 and/or β_5 are not 0.

If the variable X_3 is the exposure variable E in one's study and the variables X_1 and X_2 are confounders, then the product terms $X_1 X_3$ and $X_2 X_3$ are interaction terms for the interaction of E with X_1 and X_2, respectively. Thus, the null hypothesis that β_4 equals β_5 equals 0 is equivalent to testing no joint interaction of X_1 and X_2 with E.

The likelihood ratio statistic for comparing models 2 and 3 is then given by $-2 \ln L_2$ "hat" minus $-2 \ln L_3$ "hat," which also can be written as -2 times the natural log of the ratio of L_2 "hat" divided by L_3 "hat." This statistic has an approximate chi-square distribution in large samples. The degrees of freedom here equals 2 because there are two parameters being set equal to 0 under the null hypothesis.

When using a standard computer package to carry out this test, we must get the computer to fit the full and reduced models separately. The computer output for each model will include the log likelihood statistics of the form $-2 \ln L$ "hat." The user then simply finds the two log likelihood statistics from the output for each model being compared and subtracts one from the other to get the likelihood ratio statistic of interest.

V. The Wald Test

Focus on 1 parameter

e.g., H_0: $\beta_3 = 0$

There is another way to carry out hypothesis testing in logistic regression without using a likelihood ratio test. This second method is sometimes called **the Wald test**. This test is usually done when there is only one parameter being tested, as, for example, when comparing models 1 and 2 above.

Wald statistic (for large n):

$$Z = \frac{\hat{\beta}}{s_{\hat{\beta}}} \text{ is approximately } N(0, 1)$$

or

$$Z^2 \text{ is approximately } \chi^2 \text{ with 1 df}$$

Variable	ML Coefficient	S.E.	Chi sq	P
X_1	$\hat{\beta}_1$	$s_{\hat{\beta}_1}$	χ^2	P
•	•	•	•	•
•	•	•	•	•
•	•	•	•	•
X_j	$\hat{\beta}_j$	$s_{\hat{\beta}_j}$	χ^2	P
•	•	•	•	•
•	•	•	•	•
•	•	•	•	•
X_k	$\hat{\beta}_k$	$s_{\hat{\beta}_k}$	χ^2	P

$$\text{LR} \approx Z^2_{\text{Wald}} \text{ in large samples}$$

$$\text{LR} \neq Z^2_{\text{Wald}} \text{ in small to moderate samples}$$

LR preferred (statistical)

Wald convenient—fit only one model

EXAMPLE

Model 1: $\text{logit } P_1(\mathbf{X}) = \alpha + \beta_1 X_1 + \beta_2 X_2$

Model 2: $\text{logit } P_2(\mathbf{X}) = \alpha + \beta_1 X_1 + \beta_2 X_2 + \beta_3 X_3$

$H_0: \quad \beta_3 = 0$

$$Z = \frac{\hat{\beta}_3}{s_{\hat{\beta}_3}} \text{ is approx. } N(0, 1)$$

The Wald test statistic is computed by dividing the estimated coefficient of interest by its standard error. This test statistic has approximately a normal (0, 1), or Z, distribution in large samples. The square of this Z statistic is approximately a chi-square statistic with one degree of freedom.

In carrying out the Wald test, the information required is usually provided in the output, which lists each variable in the model followed by its ML coefficient and its standard error. Several packages also compute the chi-square statistic and a P-value.

When using the listed output, the user must find the row corresponding to the variable of interest and either compute the ratio of the estimated coefficient divided by its standard error or read off the chi-square statistic and its corresponding P-value from the output.

The likelihood ratio statistic and its corresponding squared Wald statistic give approximately the same value in very large samples; so if one's study is large enough, it will not matter which statistic is used.

Nevertheless, in small to moderate samples, the two statistics may give very different results. Statisticians have shown that the likelihood ratio statistic is better than the Wald statistic in such situations. So, when in doubt, it is recommended that the likelihood ratio statistic be used. However, the Wald statistic is somewhat convenient to use because only one model, the full model, needs to be fit.

As an example of a Wald test, consider again the comparison of models 1 and 2 described above. The Wald test for testing the null hypothesis that β_3 equals 0 is given by the Z statistic equal to β_3 "hat" divided by the standard error of β_3 "hat." The computed Z can be compared to percentage points from a standard normal table.

EXAMPLE (continued)	Or, alternatively, the Z can be squared and then compared to percentage points from a chi-square distribution with one degree of freedom.
or	
Z^2 is approximately χ^2 with one df	

Wald test for more than one parameter: requires matrices
(see *Epidemiologic Research*, Chapter 20, p. 431)

The Wald test we have just described considers a null hypothesis involving only one model parameter. There is also a Wald test that considers null hypotheses involving more than one parameter, such as when comparing models 2 and 3 above. However, this test requires knowledge of matrix theory and is beyond the scope of this presentation. The reader is referred to the text by Kleinbaum, Kupper, and Morgenstern (*Epidemiologic Research*, Chapter 20, p. 431) for a description of this test.

Third testing method:

Score statistic
(see Kleinbaum et al., *Commun. in Statist.*, 1982)

Yet another method for testing these hypotheses involves the use of a **score statistic** (see Kleinbaum et al., *Communications in Statistics*, 1982). Because this statistic is not routinely calculated by standard ML programs, and because its use gives about the same numerical chi-square values as the two techniques just presented, we will not discuss it further here.

VI. Interval Estimation: One Coefficient

Large sample confidence interval:

estimate ± (percentage point of Z × estimated standard error)

We have completed our discussion of hypothesis testing and are now ready to describe **confidence interval estimation.** We first consider interval estimation when there is only one regression coefficient of interest. The procedure typically used is to obtain a large sample confidence interval for the parameter by computing **the estimate of the parameter plus or minus a percentage point of the normal distribution times the estimated standard error.**

EXAMPLE

Model 2: $\text{logit } P_2(\mathbf{X}) = \alpha + \beta_1 X_1 + \beta_2 X_2 + \beta_3 X_3$

$100(1-\alpha)\%$ CI for β_3:

$$\hat{\beta}_3 \pm Z_{1-\frac{\alpha}{2}} \times s_{\hat{\beta}_3}$$

$\hat{\beta}_3$ and $s_{\hat{\beta}_3}$:   from printout

Z from $N(0, 1)$ tables,

e.g., $95\% \Rightarrow \alpha = 0.05$

$$\Rightarrow 1 - \frac{\alpha}{2} = 1 - 0.025 = 0.975$$

$$Z_{0.975} = 1.96$$

CI for coefficient
vs.
✓ CI for odds ratio

EXAMPLE

$\text{logit } P_2(\mathbf{X}) = \alpha + \beta_1 X_1 + \beta_2 X_2 + \beta_3 X_3$

$X_3 = (0, 1)$ variable

$$\Rightarrow \text{OR} = e^{\beta_3}$$

CI for OR: $\exp(\text{CI for } \beta_3)$

Model 2: $X_3 = (0, 1)$ exposure

$\qquad X_1$ and X_2 confounders

95% CI for OR:

$\exp(\hat{\beta}_3 \pm 1.96 s_{\hat{\beta}_3})$

Above formula assumes X_3 is coded as $(0, 1)$

As an example, if we focus on the β_3 parameter in model 2, the 100 times $(1-\alpha)\%$ confidence interval formula is given by β_3 "hat" plus or minus the corresponding $(1-\alpha/2)$th percentage point of Z times the estimated standard error of β_3 "hat."

In this formula, the values for β_3 "hat" and its standard error are found from the printout. The Z percentage point is obtained from tables of the standard normal distribution. For example, if we want a 95% confidence interval, then α is 0.05, $1-\alpha/2$ is $1-0.025$ or 0.975, and $Z_{0.975}$ is equal to 1.96.

Most epidemiologists are not interested in getting a confidence interval for the coefficient of a variable in a logistic model, but rather want a **confidence interval for an odds ratio** involving that parameter and possibly other parameters.

When only **one exposure variable** is being considered, such as X_3 in model 2, and this variable is a $(0, 1)$ variable, then the odds ratio of interest, which adjusts for the other variables in the model, is e to that parameter, for example e to β_3. In this case, the corresponding **confidence interval for the odds ratio is obtained by exponentiating the confidence limits obtained for the parameter.**

Thus, if we consider model 2, and if X_3 denotes a $(0, 1)$ exposure variable of interest and X_1 and X_2 are confounders, then a 95% confidence interval for the adjusted odds ratio e to β_3 is given by the exponential of the confidence interval for β_3, as shown here.

This formula is correct, provided that the variable X_3 is a $(0, 1)$ variable. If this variable is coded differently, such as $(-1, 1)$, or if this variable is an ordinal or interval variable, then the confidence interval formula given here must be modified to reflect the coding.

Chapter 3: Computing OR for different codings

A detailed discussion of the effect of different codings of the exposure variable on the computation of the odds ratio is described in Chapter 3 of this text. It is beyond the scope of this presentation to describe in detail the effect of different codings on the corresponding confidence interval for the odds ratio. We do, however, provide a simple example to illustrate this situation.

EXAMPLE

X_3 coded as $\begin{cases} -1 & \text{unexposed} \\ 1 & \text{exposed} \end{cases}$

$\text{OR} = \exp\left[1 - (-1)\beta_3\right] = e^{2\beta_3}$

95% CI:

$\exp\left(2\hat{\beta}_3 \pm 1.96 \times 2s_{\hat{\beta}_3}\right)$

Suppose X_3 is coded as $(-1, 1)$ instead of $(0, 1)$, so that -1 denotes unexposed persons and 1 denotes exposed persons. Then, the odds ratio expression for the effect of X_3 is given by e to 1 minus -1 times β_3, which is e to 2 times β_3. The corresponding 95% confidence interval for the odds ratio is then given by exponentiating the confidence limits for the parameter $2\beta_3$, as shown here; that is, the previous confidence interval formula is modified by multiplying β_3 and its standard error by the number 2.

VII. Interval Estimation: Interaction

No interaction: simple formula

Interaction: complex formula

The above confidence interval formulae involving a single parameter assume that there are no interaction effects in the model. When there is interaction, the confidence interval formula must be modified from what we have given so far. Because the general confidence interval formula is quite complex when there is interaction, our discussion of the modifications required will proceed by example.

EXAMPLE

Model 3: $X_3 = (0, 1)$ exposure

$\text{logit } P_3(\mathbf{X}) = \alpha + \beta_1 X_1 + \beta_2 X_2 + \beta_3 X_3$
$\qquad + \beta_4 X_1 X_3 + \beta_5 X_2 X_3$

$\widehat{\text{OR}} = \exp\left(\hat{\beta}_3 + \hat{\beta}_4 X_1 + \hat{\beta}_5 X_2\right)$

Suppose we focus on model 3, which is again shown here, and we assume that the variable X_3 is a $(0, 1)$ exposure variable of interest. Then the formula for the estimated odds ratio for the effect of X_3 controlling for the variables X_1 and X_2 is given by the exponential of the quantity β_3 "hat" plus β_4 "hat" times X_1 plus β_5 "hat" times X_2, where β_4 "hat" and β_5 "hat" are the estimated coefficients of the interaction terms $X_1 X_3$ and $X_2 X_3$ in the model.

EXAMPLE

i.e., $\widehat{OR} = e^{\hat{l}}$,

where

$$l = \beta_3 + \beta_4 X_1 + \beta_5 X_2$$

100 $(1-\alpha)$% CI for e^{l}
similar to CI formula for e^{β_3}

$$\exp\left[\hat{l} \pm Z_{1-\frac{\alpha}{2}} \sqrt{\widehat{\text{var}}(\hat{l})}\right]$$

similar to $\exp\left[\hat{\beta}_3 \pm Z_{1-\frac{\alpha}{2}} \sqrt{\widehat{\text{var}}(\hat{\beta}_3)}\right]$

$\sqrt{\widehat{\text{var}}(\bullet)}$ = standard error

General CI formula:

$$\exp\left[\hat{l} \pm Z_{1-\frac{\alpha}{2}} \sqrt{\widehat{\text{var}}(\hat{l})}\right]$$

example: $l = \beta_3 + \beta_4 X_1 + \beta_5 X_2$

General expression for l:

$$\text{ROR}_{\mathbf{X}_1, \mathbf{X}_0} = e^{\sum\limits_{i=1}^{k} \beta_i (X_{1i} - X_{0i})}$$

$\text{OR} = e^{l}$ where

$$l = \sum_{i=1}^{k} \beta_i (X_{1i} - X_{0i})$$

We can alternatively write this estimated odds ratio formula as e to the l "hat," where l is the linear function β_3 plus β_4 times X_1 plus β_5 times X_2, and l "hat" is the estimate of this linear function using the ML estimates.

To obtain a 100 times $(1-\alpha)$% confidence interval for the odds ratio e to l, we must use the linear function l the same way that we used the single parameter β_3 to get a confidence interval for β_3. The corresponding confidence interval is thus given by exponentiating the confidence interval for l.

The formula is therefore the exponential of the quantity l "hat" plus or minus a percentage point of the Z distribution times the square root of the estimated variance of l "hat." Note that the square root of the estimated variance is the standard error.

This confidence interval formula, though motivated by our example using model 3, is actually the general formula for the confidence interval for any odds ratio of interest from a logistic model. In our example, the linear function l took a specific form, but, in general, the linear function may take any form of interest.

A general expression for this linear function makes use of the general odds ratio formula described in our review. That is, the odds ratio comparing two groups identified by the vectors $\mathbf{X}_1$ and $\mathbf{X}_0$ is given by the formula e to the sum of terms of the form β_i times the difference between X_{1i} and X_{0i}, where the latter denote the values of the ith variable in each group. We can equivalently write this as e to the l, where l is the linear function given by the sum of the β_i times the difference between X_{1i} and X_{0i}. This latter formula is the general expression for l.

Interaction: variance difficult

No interaction: variance directly from printout

The difficult part in computing the confidence interval for an odds ratio involving interaction effects is the calculation for the estimated variance or corresponding square root, the standard error. When there is **no interaction,** so that the parameter of interest is a single regression coefficient, this variance is obtained directly from the variance–covariance output or from the listing of estimated coefficients and corresponding standard errors.

$$\mathrm{var}\left(\hat{l}\right) = \mathrm{var}\left[\underbrace{\sum \hat{\beta}_i \left(X_{1i} - X_{0i}\right)}_{\text{linear sum}}\right]$$

$\hat{\beta}_i$ are correlated for different i

Must use $\mathrm{var}\left(\hat{\beta}_i\right)$ and $\mathrm{cov}\left(\hat{\beta}_i,\ \hat{\beta}_j\right)$

However, when the odds ratio involves **interaction** effects, the estimated variance considers a linear sum of estimated regression coefficients. The difficulty here is that, because the coefficients in the linear sum are estimated from the same data set, these coefficients are correlated with one another. Consequently, the calculation of the estimated variance must consider both the variances and the covariances of the estimated coefficients, which makes computations somewhat cumbersome.

EXAMPLE (model 3)

$$\exp\left[\hat{l} \pm Z_{1-\frac{\alpha}{2}} \sqrt{\widehat{\mathrm{var}}\left(\hat{l}\right)}\right],$$

$$\text{where } \hat{l} = \hat{\beta}_3 + \hat{\beta}_4 X_1 + \hat{\beta}_5 X_2$$

$$\widehat{\mathrm{var}}\left(\hat{l}\right) = \widehat{\mathrm{var}}\left(\hat{\beta}_3\right) + \left(X_1\right)^2 \widehat{\mathrm{var}}\left(\hat{\beta}_4\right) + \left(X_2\right)^2 \widehat{\mathrm{var}}\left(\hat{\beta}_5\right)$$
$$+ 2X_1 \widehat{\mathrm{cov}}\left(\hat{\beta}_3,\ \hat{\beta}_4\right) + 2X_2 \widehat{\mathrm{cov}}\left(\hat{\beta}_3,\ \hat{\beta}_5\right)$$
$$+ 2X_1 X_2 \widehat{\mathrm{cov}}\left(\hat{\beta}_4,\ \hat{\beta}_5\right)$$

$\mathrm{var}\left(\hat{\beta}_i\right)$ and $\mathrm{cov}\left(\hat{\beta}_i, \hat{\beta}_j\right)$ obtained from printout BUT must specify X_1 and X_2

Returning to the interaction example, recall that the confidence interval formula is given by exponentiating the quantity l "hat" plus or minus a Z percentage point times the square root of the estimated variance of l "hat," where l "hat" is given by β_3 "hat" plus β_4 "hat" times X_1 plus β_5 "hat" times X_2.

It can be shown that the estimated variance of this linear function is given by the formula shown here.

The estimated variances and covariances in this formula are obtained from the estimated variance–covariance matrix provided by the computer output. However, the calculation of both l "hat" and the estimated variance of l "hat" requires additional specification of values for the effect modifiers in the model, which in this case are X_1 and X_2.

EXAMPLE (continued)

e.g., X_1 = AGE, X_2 = SMK:
  specification 1: $X_1 = 30, X_2 = 1$
             versus
  specification 2: $X_1 = 40, X_2 = 0$

Different specifications yield different confidence intervals

Recommendation: use "typical" or "representative" values of X_1 and X_2 e.g., $\bar{X}_1$ and $\bar{X}_2$ in quintiles

For example, if X_1 denotes AGE and X_2 denotes smoking status (SMK), then one specification of these variables is $X_1 = 30$, $X_2 = 1$, and a second specification is $X_1 = 40$, $X_2 = 0$. Different specifications of these variables will yield different confidence intervals. This should be no surprise because a model containing interaction terms implies that both the estimated odds ratios and their corresponding confidence intervals vary as the values of the effect modifiers vary.

A recommended practice is to use "typical" or "representative" values of X_1 and X_2, such as their mean values in the data, or the means of subgroups, for example, quintiles, of the data for each variable.

Computer packages do not compute

$$\widehat{var}(\hat{l})$$

Recommend: write your own program

Most standard computer packages for logistic regression do not compute the estimated variance of linear functions like l "hat" as part of the program options. Therefore, because of the cumbersome nature of the variance formula, we recommend that the user write a separate computer program for the variance calculation.

General CI formula for E, V, W model:

$$\widehat{OR} = e^{\hat{l}},$$

where

$$l = \beta + \sum_{j=1}^{p_2} \delta_j W_j$$

$$\exp\left[\hat{l} \pm Z_{1-\frac{\alpha}{2}}\sqrt{\widehat{var}(\hat{l})}\right],$$

where

$$\widehat{var}(\hat{l}) = \widehat{var}(\hat{\beta}) + \sum_{j=1}^{p_2} W_j^2 \widehat{var}(\hat{\delta}_j)$$

$$+ 2\sum_{j=1}^{p_2} W_j \widehat{cov}(\hat{\beta},\hat{\delta}_j) + 2\sum_j \sum_k W_j W_k \widehat{cov}(\hat{\delta}_j,\hat{\delta}_k)$$

Obtain $\widehat{var}$'s and $\widehat{cov}$'s from printout *but* must specify W's

For the interested reader, we provide here the general formula for the estimated variance of the linear function obtained from the E, V, W model described in the review. Recall that the estimated odds ratio for this model can be written as e to l "hat," where l is the linear function given by the sum of β plus the sum of terms of the form δ_j times W_j.

The corresponding confidence interval formula is obtained by exponentiating the confidence interval for l "hat," where the variance of l "hat" is given by the general formula shown here.

In applying this formula, the user obtains the estimated variances and covariances from the variance–covariance output. However, as in the example above, the user must specify values of interest for the effect modifiers defined by the W's in the model.

EXAMPLE

E, V, W model (model 3):
  $X_3 = E$,
  $X_1 = V_1 = W_1$
  $X_2 = V_2 = W_2$

$\hat{l} = \hat{\beta}_3 + \hat{\beta}_4 X_1 + \hat{\beta}_5 X_2$
   $= \hat{\beta} + \hat{\delta}_1 W_1 + \hat{\delta}_2 W_2$

$\beta = \beta_3$,

$p_2 = 2$, $W_1 = X_1$, $W_2 = X_2$,

       $\delta_1 = \beta_4$, and $\delta_2 = \beta_5$

Note that the example described earlier involving model 3 is a special case of the formula for the E, V, W model, with X_3 equal to E, X_1 equal to both V_1 and W_1, and X_2 equal to both V_2 and W_2. The linear function l for model 3 is shown here, both in its original form and in the E, V, W format.

To obtain the confidence interval for the model 3 example from the general formula, the following substitutions would be made in the general variance formula: $\beta = \beta_3$, $p_2 = 2$, $W_1 = X_1$, $W_2 = X_2$, $\delta_1 = \beta_4$, and $\delta_2 = \beta_5$.

VIII. Numerical Example

EVANS COUNTY, GA
$n = 609$

Before concluding this presentation, we illustrate the ML techniques described above by way of a numerical example. We consider the printout results provided below and on the following page. These results summarize the computer output for two models based on follow-up study data on a cohort of 609 white males from Evans County, Georgia.

EXAMPLE

D = CHD (0, 1)
E = CAT
C's = AGE,   CHL,   ECG, SMK, HPT
      (conts) (conts) (0, 1) (0, 1) (0, 1)

Model A Output:
$-2 \ln \hat{L} = 400.41$

	Variable	Coefficient	S.E.	Chi sq	P
	Intercept	−6.7727	1.1401	35.29	0.0000
	CAT	0.5976	0.3520	2.88	0.0896
V's	AGE	0.0322	0.0152	4.51	0.0337
	CHL	0.0087	0.0033	7.17	0.0074
	ECG	0.3695	0.2936	1.58	0.2083
	SMK	0.8347	0.3052	7.48	0.0062
	HPT	0.4393	0.2908	2.28	0.1309

unconditional ML estimation
$n = 609$, # parameters = 7

The outcome variable is coronary heart disease status, denoted as CHD, which is 1 if a person develops the disease and 0 if not. There are six independent variables of primary interest. The exposure variable is catecholamine level (CAT), which is 1 if high and 0 if low. The other independent variables are the control variables. These are denoted as AGE, CHL, ECG, SMK, and HPT.

The variable AGE is treated continuously. The variable CHL, which denotes cholesterol level, is also treated continuously. The other three variables are (0, 1) variables. ECG denotes electrocardiogram abnormality status, SMK denotes smoking status, and HPT denotes hypertension status.

EXAMPLE (continued)

Model A results are at bottom of previous page

Model B Output:

$-2 \ln \hat{L} = 347.28$

	Variable	Coefficient	S.E.	Chi sq	P
	Intercept	−4.0474	1.2549	10.40	0.0013
	CAT	−12.6809	3.1042	16.69	0.0000
	AGE	0.0349	0.0161	4.69	0.0303
	CHL	−0.0055	0.0042	1.70	0.1917
V's	ECG	0.3665	0.3278	1.25	0.2635
	SMK	0.7735	0.3272	5.59	0.0181
	HPT	1.0468	0.3316	9.96	0.0016
	CH	−2.3299	0.7422	9.85	0.0017
	CC	0.0691	0.0143	23.18	0.0000

interaction

W's

$CH = CAT \times HPT$ and $CC = CAT \times CHL$

unconditional ML estimation

$n = 609$, # parameters = 9

Model A: no interaction

$-2 \ln \hat{L} = 400.41$

Variable	Coefficient	S.E.	Chi sq	P
Intercept	−6.7727	1.1401	35.29	0.0000
CAT	0.5976	0.3520	2.88	0.0896
⋮			⋮	
HPT	0.4393	0.2908	2.28	0.1309

$\widehat{OR} = \exp(0.5976) = 1.82$

test statistic	info. available
LR	no
Wald	yes

The first set of results described by the printout information considers a model—called model A—with no interaction terms. Thus, model A contains the exposure variable CAT and the five covariables AGE, CHL, ECG, SMK and HPT. Using the E, V, W formulation, this model contains five V variables, namely, the covariables, and no W variables.

The second set of results considers a model B, which contains two interaction terms in addition to the variables contained in the first model. The two interaction terms are called CH and CC, where CH equals the product CAT × HPT and CC equals the product CAT × CHL. Thus, this model contains five V variables and two W variables, the latter being HPT and CHL.

Both sets of results have been obtained using unconditional ML estimation. Note that no matching has been done and that the number of parameters in each model is 7 and 9, respectively, which is quite small compared with the number of subjects in the data set, which is 609.

We focus for now on the set of results involving the no interaction model A. The information provided consists of the log likelihood statistic $-2 \ln L$ "hat" at the top followed by a listing of each variable and its corresponding estimated coefficient, standard error, chi-square statistic, and P-value.

For this model, because CAT is the exposure variable and there are no interaction terms, the estimated odds ratio is given by e to the estimated coefficient of CAT, which is e to the quantity 0.5976, which is 1.82. Because model A contains five V variables, we can interpret this odds ratio as an adjusted odds ratio for the effect of the CAT variable which controls for the potential confounding effects of the five V variables.

We can use this information to carry out a test of hypothesis for the significance of the estimated odds ratio from this model. Of the two test procedures described, namely, the likelihood ratio test and the Wald test, the information provided only allows us to carry out the Wald test.

EXAMPLE (continued)

LR test:

full model	reduced model
model A	model A w/o CAT

$H_0: \quad \beta = 0$

where β = coefficient of CAT in model A

reduced model (w/o CAT) printout not provided here

WALD TEST:

Variable	Coefficient	S.E.	Chi sq	P
Intercept	−6.7727	1.1401	35.29	0.0000
CAT	0.5976	0.3520	2.88	0.0896
AGE	0.0322	0.0152	4.51	0.0337
CHL	0.0087	0.0033	7.17	0.0074
ECG	0.3695	0.2936	1.58	0.2083
SMK	0.8347	0.3052	7.48	0.0062
HPT	0.4393	0.2908	2.28	0.1309

$$Z = \frac{0.5976}{0.3520} = 1.70$$

$$Z^2 = CHISQ = 2.88$$

$P = 0.0896$ misleading
(Assumes two-tailed test)
usual question: OR > 1? (one-tailed)

$$\text{one-tailed } P = \frac{\text{two-tailed } P}{2}$$

$$= \frac{0.0896}{2} = 0.0448$$

$P < 0.05 \Rightarrow$ significant at 5% level

To carry out the **likelihood ratio test,** we would need to compare two models. The full model is model A as described by the first set of results discussed here. The reduced model is a different model that contains the five covariables without the CAT variable.

The null hypothesis here is that the coefficient of the CAT variable is zero in the full model. Under this null hypothesis, the model will reduce to a model without the CAT variable in it. Because we have provided neither a printout for this reduced model nor the corresponding log likelihood statistic, we cannot carry out the likelihood ratio test here.

To carry out the **Wald test** for the significance of the CAT variable, we must use the information in the row of results provided for the CAT variable. The Wald statistic is given by the estimated coefficient divided by its standard error; from the results, the estimated coefficient is 0.5976 and the standard error is 0.3520.

Dividing the first by the second gives us the value of the Wald statistic, which is a Z, equal to 1.70. Squaring this statistic, we get the chi-square statistic equal to 2.88, as shown in the table of results.

The P-value of 0.0896 provided next to this chi square is somewhat misleading. This P-value considers a two-tailed alternative hypothesis, whereas most epidemiologists are interested in one-tailed hypotheses when testing for the significance of an exposure variable. That is, the usual question of interest is whether the odds ratio describing the effect of CAT controlling for the other variables is significantly *higher* than the null value of 1.

To obtain a one-tailed P-value from a two-tailed P-value, we simply take half of the two-tailed P-value. Thus, for our example, the one-tailed P-value is given by 0.0896 divided by 2, which is 0.0448. Because this P-value is less than 0.05, we can conclude, assuming this model is appropriate, that there is a significant effect of the CAT variable at the 5% level of significance.

EXAMPLE (continued)

H_0: $\beta = 0$
equivalent to
H_0:   adjusted OR = 1

Variable	Coefficient	S.E.	Chi sq	P
Intercept				
CAT				
AGE				
CHL	0.0087	0.0033	(7.17)	0.0074
⋮				
HPT			↑	

not of interest

95% CI for adjusted OR:
First, 95% CI for β:

$\hat{\beta} \pm 1.96 \times s_{\hat{\beta}}$

$0.5976 \pm 1.96 \times 0.3520$

CI limits for β: $\left(-0.09,\ 1.29\right)$

$\exp\left(\text{CI limits for } \beta\right)$

$= \left(e^{-0.09},\ e^{1.29}\right)$

$= \left(0.91,\ 3.63\right)$

CI contains 1,
so
do not reject H_0
at
5% level (*two-tailed*)

The Wald test we have just described tests the null hypothesis that the coefficient of the CAT variable is 0 in the model containing CAT and five covariables. An equivalent way to state this null hypothesis is that the odds ratio for the effect of CAT on CHD adjusted for the five covariables is equal to the null value of 1.

The other chi-square statistics listed in the table provide Wald tests for other variables in the model. For example, the chi-square value for the variable CHL is the squared Wald statistic that tests whether there is a significant **effect of CHL** on CHD controlling for the other five variables listed, including CAT. However, the Wald test for CHL, or for any of the other five covariables, is not of interest in this study because the only exposure variable is CAT and because the other five variables are in the model for control purposes.

A 95% confidence interval for the odds ratio for the adjusted effect of the CAT variable can be computed from the set of results for the no interaction model as follows: We first obtain a confidence interval for β, the coefficient of the CAT variable, by using the formula β "hat" plus or minus 1.96 times the standard error of β "hat." This is computed as 0.5976 plus or minus 1.96 times 0.3520. The resulting confidence limits for β are -0.09 for the lower limit and 1.29 for the upper limit.

Exponentiating the lower and upper limits gives the confidence interval for the adjusted odds ratio, which is 0.91 for the lower limit and 3.63 for the upper limit.

Note that this confidence interval contains the value 1, which indicates that a two-tailed test is not significant at the 5% level statistical significance from the Wald test. This does not contradict the earlier Wald test results, which were significant at the 5% level because using the CI, our alternative hypothesis is two-tailed instead of one-tailed.

EXAMPLE (continued)

no interaction model
   vs.
other models?

Note that the no interaction model we have been focusing on may, in fact, be inappropriate when we compare it to other models of interest. In particular, we now compare the no interaction model to the model described by the second set of printout results we have provided.

$\boxed{\text{Model B}}$ vs. Model A

We will see that this second model B, which involves interaction terms, is a better model. Consequently, the results and interpretations made about the effect of the CAT variable from the no interaction model A may be misleading.

LR test for interaction:
   $H_0: \quad \delta_1 = \delta_2 = 0$

   where δ's are coefficients of interaction terms CC and CH in model B

To compare the no interaction model with the interaction model, we need to carry out a **likelihood ratio test for the significance of the interaction terms.** The null hypothesis here is that the coefficients δ_1 and δ_2 of the two interaction terms are both equal to 0.

Full Model	Reduced Model
model B	model A
(interaction)	(no interaction)

For this test, the full model is the interaction model B and the reduced model is the no interaction model A. The likelihood ratio test statistic is then computed by taking the difference between log likelihood statistics for the two models.

$$LR = -2 \ln \hat{L}_{\text{model A}} - \left(-2 \ln \hat{L}_{\text{model B}}\right)$$

$$= 400.41 - 347.28$$

$$= 53.13$$

df = 2
   significant at .01 level

From the printout information given on pages 142–143, this difference is given by 400.41 minus 347.28, which equals 53.13. The degrees of freedom for this test is 2 because there are two parameters being set equal to 0. The chi-square statistic of 53.13 is found to be significant at the .01 level. Thus, the likelihood ratio test indicates that the interaction model is better than the no interaction model.

$\widehat{OR}$ for interaction model (B):

$$\widehat{OR} = \exp\left(\hat{\beta} + \hat{\delta}_1 CHL + \hat{\delta}_2 HPT\right)$$

$\hat{\beta} = -12.6809$ for CAT

$\hat{\delta}_1 = 0.0691$ for CC

$\hat{\delta}_2 = -2.330$ for CH

We now consider what the odds ratio is for the interaction model. As this model contains product terms CC and CH, where CC is CAT×CHL and CH is CAT×HPT, the estimated odds ratio for the effect of CAT must consider the coefficients of these terms as well as the coefficient of CAT. The formula for this estimated odds ratio is given by the exponential of the quantity β "hat" plus δ_1 "hat" times CHL plus δ_2 "hat" times HPT, where β "hat" (-12.6809) is the coefficient of CAT, δ_1 "hat" (0.0691) is the coefficient of the interaction term CC, and δ_2 "hat" (-2.330) is the coefficient of the interaction term CH.

EXAMPLE (continued)

$$\widehat{OR} = \exp[\beta + \delta_1 CHL + \delta_2 HPT]$$
$$= \exp[-12.6809 + 0.0691 CHL + (-2.3299)HPT]$$

Must specify
CHL and HPT
↑   ↑
effect modifiers

adjusted $\widehat{OR}$:

		HPT	
		0	1
	200	3.12	0.30
CHL	220	12.44	1.21
	240	49.56	4.82

CHL = 200, HPT = 0 ⇒ $\widehat{OR}$ = 3.12

CHL = 220, HPT = 1 ⇒ $\widehat{OR}$ = 1.21

$\widehat{OR}$ adjusts for AGE, CHL, ECG, SMK, and HPT (V variables)

Plugging the estimated coefficients into the odds ratio formula yields the expression: e to the quantity -12.6809 plus 0.0691 times CHL plus -2.3299 times HPT.

To obtain a numerical value from this expression, it is necessary to specify a value for CHL and a value for HPT. Different values for CHL and HPT will, therefore, yield different odds ratio values. This should be expected because the model with interaction terms should give different odds ratio estimates depending on the values of the effect modifiers, which in this case are CHL and HPT.

The table shown here illustrates different odds ratio estimates that can result from specifying different values of the effect modifiers. In this table, the values of CHL used are 200, 220, and 240; the values of HPT are 0 and 1. The cells within the table give the estimated odds ratios computed from the above expression for the odds ratio for different combinations of CHL and HPT.

For example, when CHL equals 200 and HPT equals 0, the estimated odds ratio is given by 3.12; when CHL equals 220 and HPT equals 1, the estimated odds ratio is 1.21. Each of the estimated odds ratios in this table describes the association between CAT and CHD adjusted for the five covariables AGE, CHL, ECG, SMK, and HPT because each of the covariables are contained in the model as V variables.

Confidence intervals:

$$\exp\left[\hat{l} \pm Z_{1-\frac{\alpha}{2}} \sqrt{\widehat{var}(\hat{l})}\right]$$

where

$$\hat{l} = \beta + \sum_{j=1}^{p_2} \delta_j W_j$$

To account for the variability associated with each of the odds ratios presented in the above tables, we can compute confidence intervals by using the methods we have described. The general confidence interval formula is given by e to the quantity l "hat" plus or minus a percentage point of the Z distribution times the square root of the estimated variance of l "hat," where l is the linear function shown here.

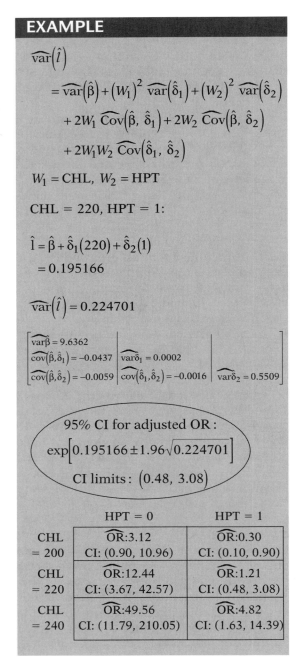

EXAMPLE

$$\widehat{\text{var}}\left(\hat{l}\right)$$

$$= \widehat{\text{var}}\left(\hat{\beta}\right) + \left(W_1\right)^2 \widehat{\text{var}}\left(\hat{\delta}_1\right) + \left(W_2\right)^2 \widehat{\text{var}}\left(\hat{\delta}_2\right)$$

$$+ 2W_1 \widehat{\text{Cov}}\left(\hat{\beta}, \hat{\delta}_1\right) + 2W_2 \widehat{\text{Cov}}\left(\hat{\beta}, \hat{\delta}_2\right)$$

$$+ 2W_1 W_2 \widehat{\text{Cov}}\left(\hat{\delta}_1, \hat{\delta}_2\right)$$

$W_1 = \text{CHL}, W_2 = \text{HPT}$

$\text{CHL} = 220, \text{HPT} = 1:$

$$\hat{l} = \hat{\beta} + \hat{\delta}_1(220) + \hat{\delta}_2(1)$$
$$= 0.195166$$

$$\widehat{\text{var}}\left(\hat{l}\right) = 0.224701$$

$$\begin{array}{lll} \widehat{\text{var}\beta} = 9.6362 & & \\ \widehat{\text{cov}}(\hat{\beta}, \hat{\delta}_1) = -0.0437 & \widehat{\text{var}\delta}_1 = 0.0002 & \\ \widehat{\text{cov}}(\hat{\beta}, \hat{\delta}_2) = -0.0059 & \widehat{\text{cov}}(\hat{\delta}_1, \hat{\delta}_2) = -0.0016 & \widehat{\text{var}\delta}_2 = 0.5509 \end{array}$$

95% CI for adjusted OR :

$$\exp\left[0.195166 \pm 1.96\sqrt{0.224701}\right]$$

CI limits : $(0.48, 3.08)$

	HPT = 0	HPT = 1
CHL = 200	$\widehat{\text{OR}}$:3.12 CI: (0.90, 10.96)	$\widehat{\text{OR}}$:0.30 CI: (0.10, 0.90)
CHL = 220	$\widehat{\text{OR}}$:12.44 CI: (3.67, 42.57)	$\widehat{\text{OR}}$:1.21 CI: (0.48, 3.08)
CHL = 240	$\widehat{\text{OR}}$:49.56 CI: (11.79, 210.05)	$\widehat{\text{OR}}$:4.82 CI: (1.63, 14.39)

For the specific interaction model (B) we have been considering, the variance of l "hat" is given by the formula shown here.

For this model, W_1 is CHL and W_2 is HPT.

As an example of a confidence interval calculation, we consider the values CHL equal to 220 and HPT equal to 1. Substituting β "hat," δ_1 "hat," and δ_2 "hat" into the formula for l "hat," we obtain the estimate l "hat" equals 0.195166.

The corresponding estimated variance is obtained by substituting into the above variance formula the estimated variances and covariances from the variance–covariance matrix. The resulting estimate of the variance of l "hat" is equal to 0.224701. The numerical values used in this calculation are shown below.

We can combine the estimates of l "hat" and its variance to obtain the 95% confidence interval. This is given by exponentiating the quantity 0.195166 plus or minus 1.96 times the square root of 0.224701. The resulting confidence limits are 0.48 for the lower limit and 3.08 for the upper limit.

The 95% confidence intervals obtained for other combinations of CHL and HPT are shown here. For example, when CHL equals 200 and HPT equals 1, the confidence limits are 0.10 and 0.90. When CHL equals 240 and HPT equals 1, the limits are 1.63 and 14.39.

EXAMPLE

Wide CIs $\Rightarrow$ estimates have
              large variances

HPT = 1:

$\widehat{OR}$ = 0.30, CI: (0.10, .90) below 1

$\widehat{OR}$ = 1.21, CI: (0.48, 3.08) includes 1

$\widehat{OR}$ = 4.82, CI: (1.63, 14.39) above 1

	$\widehat{OR}$	(two-tailed) significant?
CHL = 200:	0.30	Yes
220:	1.21	No
240:	4.82	Yes

All confidence intervals are quite wide, indicating that their corresponding point estimates have large variances. Moreover, if we focus on the three confidence intervals corresponding to HPT equal to 1, we find that the interval corresponding to the estimated odds ratio of 0.30 lies completely below the null value of 1. In contrast, the interval corresponding to the estimated odds ratio of 1.21 surrounds the null value of 1, and the interval corresponding to 4.82 lies completely above 1.

From a hypothesis testing standpoint, these results therefore indicate that the estimate of 1.21 is not statistically significant at the 5% level, whereas the other two estimates are statistically significant at the 5% level.

SUMMARY

Chapter 5: Statistical Inferences
          Using ML Techniques

This presentation is now complete. In summary, we have described two test procedures, the likelihood ratio test and the Wald test. We have also shown how to obtain interval estimates for odds ratios obtained from a logistic regression. In particular, we have described confidence interval formula for models with and without interaction terms.

We suggest that the reader review the material covered here by reading the summary outline that follows. Then you may work the practice exercises and test.

Chapter 6: Modeling Strategy Guidelines

In the next chapter, "Modeling Strategy Guidelines," we provide guidelines for determining a best model for an exposure–disease relationship that adjusts for the potential confounding and effect-modifying effects of covariables.

**Detailed
Outline**

I. **Overview** (page 128)

Focus

- testing hypotheses
- computing confidence intervals

II. **Information for making statistical inferences** (pages 128–129)

A. Maximized likelihood value: $L(\hat{\theta})$.

B. Estimated variance–covariance matrix: $\hat{V}(\hat{\theta})$ contains variances of estimated coefficients on the diagonal and covariances between coefficients off the diagonal.

C. Variable listing: contains each variable followed by ML estimate, standard error, and other information.

III. **Models for inference-making** (pages 129–130)

A. Model 1: $\text{logit } P(\mathbf{X}) = \alpha + \beta_1 X_1 + \beta_2 X_2;$

   Model 2: $\text{logit } P(\mathbf{X}) = \alpha + \beta_1 X_1 + \beta_2 X_2 + \beta_3 X_3;$

   Model 3: $\text{logit } P(\mathbf{X}) = \alpha + \beta_1 X_1 + \beta_2 X_2 + \beta_3 X_3 + \beta_4 X_1 X_3 + \beta_5 X_2 X_3.$

B. $\hat{L}_1$, $\hat{L}_2$, $\hat{L}_3$ are maximized likelihoods ($\hat{L}$) for models 1–3, respectively.

C. $\hat{L}$ is similar to R square: $\hat{L}_1 \le \hat{L}_2 \le \hat{L}_3$.

D. $-2 \ln \hat{L}_3 \le -2 \ln \hat{L}_2 \le -2 \ln \hat{L}_1$,
where $-2 \ln \hat{L}$ is called the log likelihood statistic.

IV. **The likelihood ratio (LR) test** (pages 130–134)

A. LR statistic compares two models: full (larger) model versus reduced (smaller) model.

B. H_0: some parameters in full model are equal to 0.

C. df = number of parameters in full model set equal to 0 to obtain reduced model.

D. Model 1 versus model 2: $\text{LR} = -2 \ln \hat{L}_1 - \left(-2 \ln \hat{L}_2\right)$, where H_0: $\beta_3 = 0$. This LR has approximately a chi-square distribution with one df under the null hypothesis.

E. $-2 \ln \hat{L}_1 - \left(-2 \ln \hat{L}_2\right) = -2 \ln\left(\dfrac{\hat{L}_1}{\hat{L}_2}\right)$

   where $\dfrac{\hat{L}_1}{\hat{L}_2}$ is a ratio of likelihoods.

F. How the LR test works: LR works like a chi-square statistic. For highly significant variables, LR is large and positive; for nonsignificant variables, LR is close to 0.

    G. Model 2 versus model 3: $LR = -2 \ln \hat{L}_2 - \left(-2 \ln \hat{L}_3\right)$, where $H_0: \beta_4 = \beta_5 = 0$. This LR has approximately a chi-square distribution with two df under the null hypothesis.

    H. Computer prints $-2 \ln \hat{L}$ separately for each model, so LR test requires only subtraction.

V. The Wald test (pages 134–136)

    A. Requires one parameter only to be tested, e.g., $H_0: \beta_3 = 0$.

    B. Test statistic: $Z = \dfrac{\hat{\beta}}{s_{\hat{\beta}}}$ which is approximately $N(0, 1)$ under H_0.

    C. Alternatively, Z^2 is approximately chi square with one df under H_0.

    D. LR and Z are approximately equal in large samples, but may differ in small samples.

    E. LR is preferred for statistical reasons, although Z is more convenient to compute.

    F. Example of Wald statistic for $H_0: \beta_3 = 0$ in model 2: $Z = \dfrac{\hat{\beta}_3}{s_{\hat{\beta}_3}}$.

VI. Interval estimation: one coefficient (pages 136–138)

    A. Large sample confidence interval:

        estimate $\pm$ percentage point of $Z \times$ estimated standard error.

    B. 95% CI for β_3 in model 2: $\hat{\beta}_3 \pm 1.96 s_{\hat{\beta}_3}$.

    C. If X_3 is a $(0, 1)$ exposure variable in model 2, then the 95% CI for the odds ratio of the effect of exposure adjusted for X_1 and X_2 is given by

$$\exp\left(\hat{\beta}_3 \pm 1.96 s_{\hat{\beta}_3}\right)$$

    D. If X_3 has coding other than $(0, 1)$, the CI formula must be modified.

VII. Interval estimation: interaction (pages 138–142)

    A. Model 3 example: $\widehat{OR} = e^{\hat{l}}$, where $\hat{l} = \hat{\beta}_3 + \hat{\beta}_4 X_1 + \hat{\beta}_5 X_2$

$$100(1 - \alpha)\% \text{ CI formula for OR}: \exp\left[\hat{l} \pm Z_{1-\frac{\alpha}{2}} \sqrt{\widehat{\text{var}}\left(\hat{l}\right)}\right],$$

    where

$$\widehat{\text{var}}\left(\hat{l}\right) = \widehat{\text{var}}\left(\hat{\beta}_3\right) + (X_1)^2 \, \widehat{\text{var}}\left(\hat{\beta}_4\right) + (X_2)^2 \, \widehat{\text{var}}\left(\hat{\beta}_5\right)$$
$$+ 2X_1 \, \widehat{\text{cov}}\left(\hat{\beta}_3, \hat{\beta}_4\right) + 2X_2 \, \widehat{\text{cov}}\left(\hat{\beta}_3, \hat{\beta}_5\right) + 2X_1 X_2 \, \widehat{\text{cov}}\left(\hat{\beta}_4, \hat{\beta}_5\right).$$

B. General $100(1 - \alpha)\%$ CI formula for OR:

$$\exp\left[\hat{l} \pm Z_{1-\frac{\alpha}{2}}\sqrt{\widehat{\text{var}}\left(\hat{l}\right)}\right]$$

where $\widehat{\text{OR}} = e^{\hat{l}}$,

$$\hat{l} = \sum_{i=1}^{k}\hat{\beta}_i\left(X_{1i} - X_{0i}\right) \text{ and } \text{var}\left(\hat{l}\right) = \text{var}\left(\underbrace{\Sigma\hat{\beta}_i\left(X_{1i} - X_{0i}\right)}_{\text{linear sum}}\right)$$

C. $100(1 - \alpha)\%$ CI formula for OR using E, V, W model:

$$\exp\left[\hat{l} \pm Z_{1-\frac{\alpha}{2}}\sqrt{\widehat{\text{var}}\left(\hat{l}\right)}\right],$$

where $\widehat{\text{OR}} = e^{\hat{l}}$, $\hat{l} = \hat{\beta} + \sum_{j=1}^{p_2}\hat{\delta}_j W_j$

and $\widehat{\text{var}}\left(\hat{l}\right) = \widehat{\text{var}}\left(\hat{\beta}\right) + \sum_{j=1}^{p_2}W_j^2\widehat{\text{var}}\left(\hat{\delta}_j\right) + 2\sum_{j=1}^{p_2}W_j\widehat{\text{cov}}\left(\hat{\beta},\hat{\delta}_j\right) + 2\sum_j\sum_k W_jW_k\widehat{\text{cov}}\left(\hat{\delta}_j,\hat{\delta}_k\right)$

D. Model 3 example of E, V, W model: $X_3 = E$, $X_1 = V_1$, $X_2 = V_2$, and for interaction terms, $p_2 = 2$, $X_1 = W_1$, $X_2 = W_2$.

VIII. Numerical example (pages 142–149)

A. Printout provided for two models (A and B) from Evans County, Georgia data.

B. Model A: no interaction terms; model B: interaction terms.

C. Description of LR and Wald tests for model A.

D. LR test for no interaction effect in model B: compares model B (full model) with model A (reduced model). Result: significant interaction.

E. 95% CI for OR from model B; requires use of CI formula for interaction, where $p_2 = 2$, $W_1 = $ CHL, and $W_2 = $ HPT.

Practice Exercises

A prevalence study of predictors of surgical wound infection in 265 hospitals throughout Australia collected data on 12,742 surgical patients (McLaws et al., *Med. J. of Australia*, 1988). For each patient, the following independent variables were determined: type of hospital (public or private), size of hospital (large or small), degree of contamination of surgical site (clean or contaminated), and age and sex of the patient. A logistic model was fit to this data to predict whether or not the patient developed a surgical wound infection during hospitalization. The largest model fit included all of the above variables and all possible two-way interaction terms. The abbreviated variable names and the manner in which the variables were coded in the model are described as follows:

Variable	Abbreviation	Coding
Type of hospital	HT	1 = public, 0 = private
Size of hospital	HS	1 = large, 0 = small
Degree of contamination	CT	1 = contaminated, 0 = clean
Age	AGE	continuous
Sex	SEX	1 = female, 0 = male

1. State the logit form of a no interaction model that includes all of the above predictor variables.

2. State the logit form of a model that extends the model of Exercise 1 by adding all possible pairwise products of different variables.

3. Suppose you want to carry out a (global) test for whether any of the two-way product terms (considered collectively) in your interaction model of Exercise 2 are significant. State the null hypothesis, the form of the appropriate (likelihood ratio) test statistic, and the distribution and degrees of freedom of the test statistic under the null hypothesis of no interaction effects in your model of Exercise 2.

Suppose the test for interaction in Exercise 3 is nonsignificant, so that you felt justified to drop all pairwise products from your model. The remaining model will, therefore, contain only those variables given in the above listing.

4. Consider a test for the effect of hospital type (HT) adjusted for the other variables in the no interaction model. Describe the likelihood ratio test for this effect by stating the following: the null hypothesis, the formula for the test statistic, and the distribution and degrees of freedom of the test statistic under the null hypothesis.

5. For the same question as described in Exercise 4, that is, concerning the effect of HT controlling for the other variables in the model, describe the Wald test for this effect by providing the null hypothesis, the formula for the test statistic, and the distribution of the test statistic under the null hypothesis.

6. Based on the study description preceding Exercise 1, do you think that the likelihood ratio and Wald test results will be approximately the same? Explain.

7. Give a formula for a 95% confidence interval for the odds ratio describing the effect of HT controlling for the other variables in the no interaction model.

(**Note:** In answering all of the above questions, make sure to state your answers in terms of the coefficients and variables that you specified in your answers to Exercises 1 and 2.)

Consider the following printout results that summarize the computer output for two models based on follow-up study data on 609 white males from Evans County, Georgia:

Model I OUTPUT:

$-2 \ln \hat{L} = 400.41$

Variable	Coefficient	S.E.	Chi sq	P
Intercept	−6.7727	1.1401	35.29	0.0000
CAT	0.5976	0.3520	2.88	0.0896
AGE	0.0322	0.0152	4.51	0.0337
CHL	0.0087	0.0033	7.17	0.0074
ECG	0.3695	0.2936	1.58	0.2083
SMK	0.8347	0.3052	7.48	0.0062
HPT	0.4393	0.2908	2.28	0.1309

Model II OUTPUT:

$-2 \ln \hat{L} = 357.09$

Variable	Coefficient	S.E.	Chi sq	P
Intercept	−3.9311	1.2502	9.89	0.0017
CAT	−14.0717	3.1219	20.32	0.0000
AGE	0.0323	0.0162	3.95	0.0468
CHL	−0.0045	0.0041	1.16	0.2813
ECG	0.3573	0.3263	1.20	0.2735
SMK	0.8066	0.3264	6.11	0.0135
HPT	0.6070	0.3025	4.03	0.0448
CC = CAT×CHL	0.0683	0.0143	22.74	0.0000

In the above models, the variables are coded as follows: CAT(1 = high, 0 = low), AGE(continuous), CHL(continuous), ECG(1 = abnormal, 0 = normal), SMK(1 = ever, 0 = never), HPT(1 = hypertensive, 0 = normal). The outcome variable is CHD status(1 = CHD, 0 = no CHD).

8. For model I, test the hypothesis for the effect of CAT on the development of CHD. State the null hypothesis in terms of an odds ratio parameter, give the formula for the test statistic, state the distribution of the test statistic under the null hypothesis, and, finally, carry out the test using the above printout for model I. Is the test significant?

9. Using the printout for model I, compute the point estimate and a 95% confidence interval for the odds ratio for the effect of CAT on CHD controlling for the other variables in the model.

10. Now consider model II: Carry out the likelihood ratio test for the effect of the product term CC on the outcome, controlling for the other variables in the model. Make sure to state the null hypothesis in terms of a model coefficient, give the formula for the test statistic and its distribution and degrees of freedom under the null hypothesis, and report the P-value. Is the test result significant?

11. Carry out the Wald test for the effect of CC on outcome, controlling for the other variables in model II. In carrying out this test, provide the same information as requested in Exercise 10. Is the test result significant? How does it compare to your results in Exercise 10? Based on your results, which model is more appropriate, model I or II?

12. Using the output for model II, give a formula for the point estimate of the odds ratio for the effect of CAT on CHD, which adjusts for the confounding effects of AGE, CHL, ECG, SMK, and HPT and allows for the interaction of CAT with CHL.

13. Use the formula for the adjusted odds ratio in Exercise 12 to compute numerical values for the estimated odds ratio for the following cholesterol values: CHL = 220 and CHL = 240.

14. Give a formula for the 95% confidence interval for the adjusted odds ratio described in Exercise 12 when CHL = 220. In stating this formula, make sure to give an expression for the estimated variance portion of the formula in terms of variances and covariances obtained from the variance–covariance matrix.

Test

The following printout provides information for the fitting of two logistic models based on data obtained from a matched case-control study of cervical cancer in 313 women from Sydney, Australia (Brock et al., *J. Nat. Cancer Inst.*, 1988). The outcome variable is cervical cancer status (1 = present, 0 = absent). The matching variables are age and socioeconomic status. Additional independent variables not matched on are smoking status, number of lifetime sexual partners, and age at first sexual intercourse. The independent variables not involved in the matching are listed below, together with their computer abbreviation and coding scheme.

Variable	Abbreviation	Coding
Smoking status	SMK	1 = ever, 0 = never
Number of sexual partners	NS	1 = 4+, 0 = 0–3
Age at first intercourse	AS	1 = 20+, 0 = ≤ 19

PRINTOUT:

Model I

$-2 \ln \hat{L} = 174.97$

Variable	β	S.E.	Chi sq	P
SMK	1.4361	0.3167	20.56	0.0000
NS	0.9598	0.3057	9.86	0.0017
AS	−0.6064	0.3341	3.29	0.0695

Model II

$-2 \ln \hat{L} = 171.46$

Variable	β	S.E.	Chi sq	P
SMK	1.9381	0.4312	20.20	0.0000
NS	1.4963	0.4372	11.71	0.0006
AS	−0.6811	0.3473	3.85	0.0499
SMK×NS	−1.1128	0.5997	3.44	0.0635

Variance–Covariance Matrix (Model II)

	SMK	NS	AS	SMK $\times$ NS
SMK	0.1859			
NS	0.1008	0.1911		
AS	-0.0026	-0.0069	0.1206	
SMK×NS	-0.1746	-0.1857	0.0287	0.3596

1. What method of estimation was used to obtain estimates of parameters for both models, conditional or unconditional ML estimation? Explain.

2. Why are the variables age and socioeconomic status missing from the printout, even though these were variables matched on in the study design?

3. For model I, test the hypothesis for the effect of SMK on cervical cancer status. State the null hypothesis in terms of an odds ratio parameter, give the formula for the test statistic, state the distribution of the test statistic under the null hypothesis, and, finally, carry out the test using the above printout for model I. Is the test significant?

4. Using the printout for model I, compute the point estimate and 95% confidence interval for the odds ratio for the effect of SMK controlling for the other variables in the model.

5. Now consider model II: Carry out the likelihood ratio test for the effect of the product term SMK $\times$ NS on the outcome, controlling for the other variables in the model. Make sure to state the null hypothesis in terms of a model coefficient, give the formula for the test statistic and its distribution and degrees of freedom under the null hypothesis, and report the P-value. Is the test significant?

6. Carry out the Wald test for the effect of SMK $\times$ NS, controlling for the other variables in model II. In carrying out this test, provide the same information as requested in Question 3. Is the test significant? How does it compare to your results in Question 5?

7. Using the output for model II, give a formula for the point estimate of the odds ratio for the effect of SMK on cervical cancer status, which adjusts for the confounding effects of NS and AS and allows for the interaction of NS with SMK.

8. Use the formula for the adjusted odds ratio in Question 7 to compute numerical values for the estimated odds ratios when NS = 1 and when NS = 0.

9. Give a formula for the 95% confidence interval for the adjusted odds ratio described in Question 8 (when NS = 1). In stating this formula, make sure to give an expression for the estimated variance portion of the formula in terms of variances and covariances obtained from the variance–covariance matrix.

10. Use your answer to Question 9 and the estimated variance–covariance matrix to carry out the computation of the 95% confidence interval described in Question 7.

11. Based on your answers to the above questions, which model, point estimate, and confidence interval for the effect of SMK on cervical cancer status are more appropriate, those computed for model I or those computed for model II? Explain.

Answers to Practice Exercises

1. logit $P(\mathbf{X}) = \alpha + \beta_1 HT + \beta_2 HS + \beta_3 CT + \beta_4 AGE + \beta_5 SEX$.

2. logit $P(\mathbf{X}) = \alpha + \beta_1 HT + \beta_2 HS + \beta_3 CT + \beta_4 AGE + \beta_5 SEX$
   $+ \beta_6 HT \times HS + \beta_7 HT \times CT + \beta_8 HT \times AGE + \beta_9 HT \times SEX$
   $+ \beta_{10} HS \times CT + \beta_{11} HS \times AGE + \beta_{12} HS \times SEX$
   $+ \beta_{13} CT \times AGE + \beta_{14} CT \times SEX + \beta_{15} AGE \times SEX$.

3. $H_0: \beta_6 = \beta_7 = \ldots = \beta_{15} = 0$, i.e., the coefficients of all product terms are zero;

   likelihood ratio statistic: LR $= -2 \ln \hat{L}_1 - (-2 \ln \hat{L}_2)$, where $\hat{L}_1$ is the maximized likelihood for the reduced model (i.e., Exercise 1 model) and $\hat{L}_2$ is the maximized likelihood for the full model (i.e., Exercise 2 model);

   distribution of LR statistic: chi square with 10 degrees of freedom.

4. $H_0: \beta_1 = 0$, where β_1 is the coefficient of HT in the no interaction model; alternatively, this null hypothesis can be stated as $H_0: OR = 1$, where OR denotes the odds ratio for the effect of HT adjusted for the other four variables in the no interaction model.

   likelihood ratio statistic: LR $= -2 \ln \hat{L}_0 - (-2 \ln \hat{L}_1)$, where $\hat{L}_0$ is the maximized likelihood for the reduced model (i.e., Exercise 1 model less the HT term and its corresponding coefficient) and $\hat{L}_1$ is the maximized likelihood for the full model (i.e., Exercise 1 model);

   distribution of LR statistic: approximately chi square with one degree of freedom.

5.   The null hypothesis for the Wald test is the same as that given for the likelihood ratio test in Exercise 4. H_0: $\beta_1 = 0$ or, equivalently, H_0: OR = 1, where OR denotes the odds ratio for the effect of HT adjusted for the other four variables in the no interaction model;

Wald test statistic: $Z = \dfrac{\hat{\beta}_1}{s_{\hat{\beta}_1}}$, where β_1 is the coefficient of HT in the no interaction model;

distribution of Wald statistic: approximately normal $(0, 1)$ under H_0; alternatively, the square of the Wald statistic, i.e., Z^2, is approximately chi square with one degree of freedom.

6.   The sample size for this study is 12,742, which is very large; consequently, the Wald and LR test statistics should be approximately the same.

7.   The odds ratio of interest is given by e_{β_1}, where β_1 is the coefficient of HT in the no interaction model; a 95% confidence interval for this odds ratio is given by the following formula:

$$\exp\left[\hat{\beta}_1 \pm 1.96\sqrt{\widehat{\mathrm{var}}(\hat{\beta}_1)}\,\right],$$

where $\widehat{\mathrm{var}}(\hat{\beta}_1)$ is obtained from the variance–covariance matrix or, alternatively, by squaring the value of the standard error for $\hat{\beta}_1$ provided by the computer in the listing of variables and their estimated coefficients and standard errors.

8. H_0: $\beta_{CAT} = 0$ in the no interaction model (model I), or alternatively, H_0: OR = 1, where OR denotes the odds ratio for the effect of CAT on CHD status, adjusted for the five other variables in model I;

test statistic: Wald statistic $Z = \dfrac{\hat{\beta}_{CAT}}{s_{\hat{\beta}_{CAT}}}$, which is approximately normal

$(0, 1)$ under H_0, or alternatively,

Z^2 is approximately chi square with one degree of freedom under H_0;

test computation: $Z = \dfrac{0.5976}{0.3520} = 1.69$; alternatively, $Z^2 = 2.88$;

the one-tailed P-value is $0.0896/2 = 0.0448$, which is significant at the 5% level.

9.   The point estimate of the odds ratio for the effect of CAT on CHD adjusted for the other variables in model I is given by $e^{0.5976} = 1.82$. The 95% interval estimate for the above odds ratio is given by

$$\exp\left[\hat{\beta}_{CAT} \pm 1.96\sqrt{\widehat{\mathrm{var}}(\hat{\beta}_{CAT})}\,\right] = (0.5976 \pm 1.96 \times 0.3520)$$

$$= \exp(0.5976 \pm 0.6899)$$

$$= \left(e^{-0.0923}, e^{1.2875}\right) = (0.91,\ 3.62).$$

10. The null hypothesis for the likelihood ratio test for the effect of CC: H_0: $\beta_{CC} = 0$ where β_{CC} is the coefficient of CC in model II. Likelihood ratio statistic: LR $= -2 \ln \hat{L}_I - (-2 \ln \hat{L}_{II})$ where $\hat{L}_I$ and $\hat{L}_{II}$ are the maximized likelihood functions for models I and II, respectively. This statistic has approximately a chi-square distribution with one degree of freedom under the null hypothesis.

    Test computation: LR $= 400.4 - 357.0 = 43.4$. The P-value is 0.0000 to four decimal places. Because P is very small, the null hypothesis is rejected and it is concluded that there is a significant effect of the CC variable, i.e., there is significant interaction of CHL with CAT.

11. The null hypothesis for the Wald test for the effect of CC is the same as that for the likelihood ratio test: H_0: $\beta_{CC} = 0$, where β_{CC} is the coefficient of CC in model II.

    Wald statistic: $Z = \dfrac{\hat{\beta}_{CC}}{s\hat{\beta}_{CC}}$, which is approximately normal (0, 1) under H_0, or alternatively, Z^2 is approximately chi square with one degree of freedom under H_0;

    test computation: $Z = \dfrac{0.0683}{0.0143} = 4.77$; alternatively, $Z^2 = 22.74$;

    the two-tailed P-value is 0.0000, which is very significant.

    The LR statistic is 43.4, which is almost twice as large as the square of the Wald statistic; however, both statistics are very significant, resulting in the same conclusion of rejecting the null hypothesis.

    Model II is more appropriate than model I because the test for interaction is significant.

12. The formula for the estimated odds ratio is given by

    $$\widehat{OR}_{adj} = \exp\left(\hat{\beta}_{CAT} + \hat{\delta}_{CC}CHL\right) = \exp\left(-14.0717 + 0.0683\ CHL\right),$$

    where the coefficients come from model II and the confounding effects of AGE, CHL, ECG, SMK, and HPT are adjusted.

13. Using the adjusted odds ratio formula given in Exercise 12, the estimated odds ratio values for CHL equal to 220 and 240 are

    CHL = 220: $\exp[-14.0707 + 0.0683(220)] = \exp(0.9553) = 2.60$;
    CHL = 240: $\exp[-14.0707 + 0.0683(240)] = \exp(2.3213) = 10.19$.

14. Formula for the 95% confidence interval for the adjusted odds ratio when CHL = 220:

    $$\exp\left[\hat{l} \pm 1.96\sqrt{\widehat{\text{var}}(\hat{l})}\right], \text{ where } \hat{l} = \hat{\beta}_{CAT} + \hat{\delta}_{CC}(220)$$

    and $\widehat{\text{var}}(\hat{l}) = \widehat{\text{var}}(\hat{\beta}_{CAT}) + (220)^2\ \widehat{\text{var}}(\hat{\delta}_{CC}) + 2(220)\ \widehat{\text{cov}}(\hat{\beta}_{CAT}, \hat{\delta}_{CC})$

    where $\widehat{\text{var}}(\hat{\beta}_{CAT})$, $\widehat{\text{var}}(\hat{\delta}_{CC})$, and $\widehat{\text{cov}}(\hat{\beta}_{CAT}, \hat{\delta}_{CC})$ are obtained from the printout of the variance–covariance matrix.

6

Modeling Strategy Guidelines

Introduction

We begin this chapter by giving the rationale for having a strategy to determine a "best" model. Focus is on a logistic model containing a single dichotomous exposure variable which adjusts for potential confounding and potential interaction effects of covariates considered for control. A strategy is recommended which has three stages: (1) variable specification, (2) interaction assessment, and (3) confounding assessment followed by consideration of precision. The initial model has to be "hierarchically well formulated," a term to be defined and illustrated. Given an initial model, we recommend a strategy involving a "hierarchical backward elimination procedure" for removing variables. In carrying out this strategy, statistical testing is allowed for assessing interaction terms but is not allowed for assessing confounding. Further description of interaction and confounding assessment is given in the next chapter (Chapter 7).

Abbreviated Outline

The outline below gives the user a preview of the material in this chapter. A detailed outline for review purposes follows the presentation.

Objectives

Upon completion of this chapter, the learner should be able to:

1. State and recognize the three stages of the recommended modeling strategy.
2. Define and recognize a hierarchically well-formulated logistic model.
3. State, recognize, and apply the recommended strategy for choosing potential confounders in one's model.
4. State, recognize, and apply the recommended strategy for choosing potential effect modifiers in one's model.
5. State and recognize the rationale for a hierarchically well-formulated model.
6. State and apply the hierarchical backward elimination strategy.
7. State and apply the Hierarchy Principle.
8. State whether or not significance testing is allowed for the assessment of interaction and/or confounding.

Presentation

I. Overview

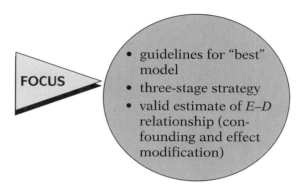

FOCUS

- guidelines for "best" model
- three-stage strategy
- valid estimate of E–D relationship (confounding and effect modification)

This presentation gives guidelines for determining the "best" model when carrying out mathematical modeling using logistic regression. We focus on a strategy involving three stages. The goal of this strategy is to obtain a valid estimate of an exposure–disease relationship that accounts for confounding and effect modification.

II. Rationale for a Modeling Strategy

We begin by explaining the rationale for a modeling strategy.

Minimum information in most study reports
e.g., little explanation about strategy

Most epidemiologic research studies in the literature, regardless of the exposure–disease question of interest, provide a minimum of information about modeling methods used in the data analysis. Typically, only the final results from modeling are reported, with little accompanying explanation about the strategy used in obtaining such results.

Information often *not* provided:
- how variables chosen
- how variables selected
- how effect modifiers assessed
- how confounders assessed

For example, information is often *not* provided as to how variables are chosen for the initial model, how variables are selected for the final model, and how effect modifiers and confounders are assessed for their role in the final model.

Guidelines needed
- to assess validity of results
- to help researchers know what information to provide
- to encourage consistency in strategy
- for a variety of modeling procedures

Without meaningful information about the modeling strategy used, it is difficult to assess the validity of the results provided. Thus, there is need for guidelines regarding modeling strategy to help researchers know what information to provide.

In practice, most modeling strategies are ad hoc; in other words, researchers often make up a strategy as they go along in their analysis. The general guidelines that we recommend here encourage more consistency in the strategy used by different researchers.

Guidelines applicable to
  logistic regression,
  multiple linear regression
  Cox PH regression

Modeling strategy guidelines are also important for modeling procedures other than logistic regression. In particular, classical multiple linear regression and Cox proportional hazards regression, although having differing model forms, all have in common with logistic regression the goal of describing exposure–disease relationships when used in epidemiologic research. The strategy offered here, although described in the context of logistic regression, is applicable to a variety of modeling procedures.

Two modeling goals
(1) to obtain a valid E–D estimate
(2) to obtain a good predictive model

(different strategies for different goals)

There are typically two goals of mathematical modeling: One is to obtain a valid estimate of an exposure–disease relationship and the other is to obtain a good predictive model. Depending on which of these is the primary goal of the researcher, different strategies for obtaining the "best" model are required.

Prediction goal:
  use computer algorithms

When the goal is "prediction," it is appropriate to use computer algorithms, such as backward elimination or all possible regressions, which are built into computer packages for different models.

Validity goal
• our focus
• for etiologic research
• standard computer algorithms not appropriate

Our focus in this presentation is on the goal of obtaining a valid measure of effect. This goal is characteristic of most etiologic research in epidemiology. For this goal, standard computer algorithms do not apply because the roles that variables—such as confounders and effect modifiers—play in the model must be given special attention.

III. Overview of Recommended Strategy

Three stages
(1) variable specification
(2) interaction assessment
(3) confounding assessment followed by precision

The modeling strategy we recommend involves three stages: (1) **variable specification, (2) interaction assessment,** and (3) **confounding assessment followed by consideration of precision.** We have listed these stages in the order that they should be addressed.

Variable specification
• restricts attention to clinically or biologically meaningful variables
• provides largest possible initial model

Variable specification is addressed first because this step allows the investigator to use the research literature to restrict attention to clinically or biologically meaningful independent variables of interest. These variables can then be defined in the model to provide the largest possible meaningful model to be initially considered.

Interaction prior to confounding
- if strong interaction, then confounding irrelevant

Interaction assessment is carried out next, prior to the assessment of confounding. The reason for this ordering is that if there is strong evidence of interaction involving certain variables, then the assessment of confounding involving these variables becomes irrelevant.

EXAMPLE

Suppose *gender* is effect modifier for E–D relationship:

$\widehat{OR}$ males = 5.4, $\widehat{OR}$ females = 1.2

interaction

Overall average = 3.5
   not appropriate

Misleading because of separate effects for males and females

For example, suppose we are assessing the effect of an exposure variable E on some disease D, and we find strong evidence that **gender** is an effect modifier of the E–D relationship. In particular, suppose that the odds ratio for the effect of E on D is 5.4 for males but only 1.2 for females. In other words, the data indicate that the E–D relationship is different for males than for females, that is, there is interaction due to gender.

For this situation, it would *not* be appropriate to combine the two odds ratio estimates for males and females into a single overall adjusted estimate, say 3.5, that represents an "average" of the male and female odds ratios. Such an overall "average" is used to control for the confounding effect of gender in the absence of interaction; however, if interaction is present, the use of a single adjusted estimate is a misleading statistic because it masks the finding of a separate effect for males and females.

Assess interaction before confounding

Thus, we recommend that if one wishes to assess interaction and also consider confounding, then the assessment of interaction comes first.

Interaction may not be of interest:
- skip interaction stage
- proceed directly to confounding

However, the circumstances of the study may indicate that the assessment of interaction is not of interest or is biologically unimportant. In such situations, the interaction stage of the strategy can then be skipped, and one proceeds directly to the assessment of confounding.

EXAMPLE

Study goal: single overall estimate. Then interaction not appropriate

For example, the goal of a study may be to obtain a **single** overall estimate of the effect of an exposure adjusted for several factors, regardless of whether or not there is interaction involving these factors. In such a case, then, interaction assessment is not appropriate.

If interaction present

- do not assess confounding for effect modifiers
- assessing confounding for other variables difficult and subjective

On the other hand, if interaction assessment is considered worthwhile, and, moreover, if significant interaction is found, then this precludes assessing confounding for those variables identified as effect modifiers. Also, as we will describe in more detail later, assessing confounding for variables other than effect modifiers can be quite difficult and, in particular, extremely subjective, when interaction is present.

Confounding followed by precision:

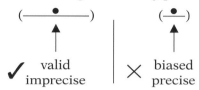

The final stage of our strategy calls for the assessment of confounding followed by consideration of **precision.** This means that it is more important to get a valid point estimate of the *E–D* relationship that controls for confounding than to get a narrow confidence interval around a biased estimate that does not control for confounding.

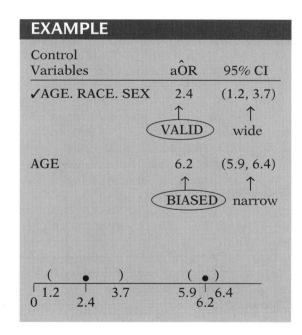

EXAMPLE

Control Variables	a$\hat{\text{OR}}$	95% CI
✓AGE. RACE. SEX	2.4	(1.2, 3.7)
AGE	6.2	(5.9, 6.4)

For example, suppose controlling for **AGE, RACE,** and **SEX** simultaneously gave an adjusted odds ratio estimate of 2.4 with a 95% confidence interval ranging between 1.2 and 3.7, whereas controlling for **AGE alone** gave an odds ratio of 6.2 with a 95% confidence interval ranging between 5.9 and 6.4.

Then, assuming that **AGE, RACE,** and **SEX** are **considered important risk factors** for the disease of interest, we would prefer to use the odds ratio of 2.4 over the odds ratio of 6.2. This is because the 2.4 value results from controlling for all the relevant variables and, thus, gives us a more valid answer than the value of 6.2, which controls for only one of the variables.

Thus, even though there is a much narrower confidence interval around the 6.2 estimate than around the 2.4, the gain in precision from using 6.2 does not offset the bias in this estimate when compared to the more valid 2.4 value.

VALIDITY BEFORE PRECISION

↓	↓
✓ right answer	precise answer

In essence, then, **validity takes precedence over precision, so that it is more important to get the right answer than a precise answer.** Thus, in the third stage of our strategy, we seek an estimate that controls for confounding and is, over and above this, as precise as possible.

Confounding : *no statistical testing*

↓

Validity—systematic error

(Statistical testing—random error)

When later describing this last stage in more detail we will emphasize that **the assessment of confounding is carried out without using statistical testing.** This follows from general epidemiologic principles in that confounding is a validity issue which addresses systematic rather than random error. Statistical testing is appropriate for considering random error rather than systematic error.

Confounding in logistic regression—a validity issue

Computer algorithms no good (involve statistical testing)

Our suggestions for assessing confounding using logistic regression are consistent with the principle that confounding is a validity issue. Standard computer algorithms for variable selection, such as forward inclusion or backward elimination procedures, are not appropriate for assessing confounding because they involve statistical testing.

Statistical issues beyond scope of this presentation:
- multicollinearity
- multiple testing
- influential observations

Multicollinearity
- independent variables approximately determined by other independent variables
- regression coefficients unreliable

Before concluding this overview section, we point out a few statistical issues needing attention but which are beyond the scope of this presentation. These issues are **multicollinearity, multiple testing,** and **influential observations.**

Multicollinearity occurs when one or more of the independent variables in the model can be approximately determined by some of the other independent variables. When there is multicollinearity, the estimated regression coefficients of the fitted model can be highly unreliable. Consequently, any modeling strategy must check for possible multicollinearity at various steps in the variable selection process.

Multiple testing
- the more tests, the more likely significant findings, even if no real effects
- variable selection procedures may yield an incorrect model because of multiple testing

Multiple testing occurs from the many tests of significance that are typically carried out when selecting or eliminating variables in one's model. The problem with doing several tests on the same data set is that the more tests one does, the more likely one can obtain statistically significant results even if there are no real associations in the data. Thus, the process of variable selection may yield an incorrect model because of the number of tests carried out. Unfortunately, there is no foolproof method for adjusting for multiple testing, even though there are a few rough approaches available.

Influential observations
- individual data may influence regression coefficients, e.g., outlier
- coefficients may change if outlier is dropped from analysis

Influential observations refer to data on individuals that may have a large influence on the estimated regression coefficients. For example, an outlier in one or more of the independent variables may greatly affect one's results. If a person with an outlier is dropped from the data, the estimated regression coefficients may greatly change from the coefficients obtained when that person is retained in the data. Methods for assessing the possibility of influential observations should be considered when determining a best model.

IV. Variable Specification Stage

- define clinically or biologically meaningful independent variables
- provide initial model

Specify D, E, C_1, C_2, . . . , C_p based on
- study goals
- literature review
- theory

Specify V's based on
- prior research or theory
- possible statistical problems

At the variable specification stage, clinically or biologically meaningful independent variables are defined in the model to provide the largest model to be initially considered.

We begin by specifying the D and E variables of interest together with the set of risk factors C_1 through C_p to be considered for control. These variables are defined and measured by the investigator based on the goals of one's study and a review of the literature and/or biological theory relating to the study.

Next, we must specify the V's, which are functions of the C's that go into the model as potential confounders. Generally, we recommend that the choice of V's be based primarily on prior research or theory, with some consideration of possible statistical problems like multicollinearity that might result from certain choices.

For example, if the C's are AGE, RACE, and SEX, one choice for the V's is the C's themselves. Another choice includes AGE, RACE, and SEX plus more complicated functions such as AGE^2, AGE × RACE, RACE × SEX, and AGE × SEX.

We would recommend any of the latter four variables only if prior research or theory supported their inclusion in the model. Moreover, even if biologically relevant, such variables may be omitted from consideration if multicollinearity is found.

EXAMPLE

C's: AGE, RACE, SEX

V's:

Choice 1: AGE, RACE, SEX

Choice 2: AGE, RACE, SEX, AGE^2,
           AGE × RACE, RACE × SEX,
           AGE × SEX

✓ Simplest choice for V's

  the C's themselves (or a subset of C's)

The simplest choice for the V's is the C's themselves. If the number of C's is very large, it may even be appropriate to consider a smaller subset of the C's considered to be most relevant and interpretable based on prior knowledge.

Specify W's: (in model as $E \times W$)

  restrict W's to be V's themselves or products of two V's

  (i.e., in model as $E \times V_i$ and $E \times V_i \times V_j$)

Once the V's are chosen, the next step is to determine the W's. These are the effect modifiers that go into the model as product terms with E, that is, these variables are of the form E times W.

We recommend that the choice of W's be restricted either to the V's themselves or to product terms involving two V's. Correspondingly, the product terms in the model are recommended to be of the form E times V and E times V_i times V_j, where V_i and V_j are two distinct V's.

Most situations:

  specify V's and W's as C's or subset of C's

For most situations, we recommend that both the V's and the W's be the C's themselves, or even a subset of the C's.

EXAMPLE

$C_1, C_2, C_3, =$ AGE, RACE, SEX

$V_1, V_2, V_3, =$ AGE, RACE, SEX

W's = subset of AGE, RACE, SEX

As an example, if the C's are AGE, RACE, and SEX, then a simple choice would have the V's be AGE, RACE, and SEX and the W's be a subset of AGE, RACE, and SEX thought to be biologically meaningful as effect modifiers.

Rationale for W's (common sense):

  Product terms more complicated than EV_iV_j are

• difficult to interpret
• typically cause multicollinearity
✓ Simplest choice: use EV_i terms only

The **rationale** for our recommendation about the W's is based on the following commonsense considerations:

• product terms more complicated than EV_iV_j are usually **difficult to interpret** even if found significant; in fact, even terms of the form EV_iV_j are often uninterpretable.
• product terms more complicated than EV_iV_j typically will cause **multicollinearity** problems; this is also likely for EV_iV_j terms, so the simplest way to reduce the potential for multicollinearity is to use EV_i terms only.

Variable Specification Summary Flow Diagram

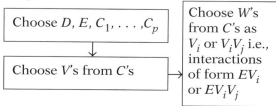

In summary, at the variable specification stage, the investigator defines the largest possible model initially to be considered. The flow diagram at the left shows first the choice of D, E, and the C's, then the choice of the V's from the C's and, finally, the choice of the W's in terms of the C's.

V. Hierarchically Well-Formulated Models

Initial model structure: HWF

When choosing the V and W variables to be included in the initial model, the investigator must ensure that the model has a certain structure to avoid possibly misleading results. This structure is called a **hierarchically well-formulated model,** abbreviated as HWF, which we define and illustrate in this section.

Model contains **all lower-order components**

A hierarchically well-formulated model is a model satisfying the following characteristic: Given any variable in the model, all lower-order components of the variable must also be contained in the model.

EXAMPLE

Not HWF model:

$$\text{logit } P(\mathbf{X}) = \alpha + \beta E + \gamma_1 V_1 + \gamma_2 V_2$$
$$+ \delta_1 EV_1 + \delta_2 EV_2 + \delta_3 EV_1 V_2$$

Components of $EV_1 V_2$:

$E, V_1, V_2, EV_1, EV_2, V_1 V_2$

$\uparrow$ not in model

To understand this definition, let us look at an example of a model that is *not* hierarchically well formulated. Consider the model given in logit form as logit $P(\mathbf{X})$ equals α plus βE plus $\gamma_1 V_1$ plus $\gamma_2 V_2$ plus the product terms $\delta_1 EV_1$ plus $\delta_2 EV_2$ plus $\delta_3 EV_1 V_2$.

For this model, let us focus on the three-factor product term $EV_1 V_2$. This term has the following lower-order components: E, V_1, V_2, EV_1, EV_2, and $V_1 V_2$. Note that the last component $V_1 V_2$ is not contained in the model. Thus, the model is not hierarchically well formulated.

EXAMPLE

HWF model:

$$\text{logit } P(\mathbf{X}) = \alpha + \beta E + \gamma_1 V_1 + \gamma_2 V_2$$
$$+ \delta_1 EV_1 + \delta_2 EV_2$$

Components of EV_1:

E, V_1 both in model

In contrast, the model given by logit $P(\mathbf{X})$ equals α plus βE plus $\gamma_1 V_1$ plus $\gamma_2 V_2$ plus the product terms $\delta_1 EV_1$ plus $\delta_2 EV_2$ is hierarchically well formulated because the lower-order components of each variable in the model are also in the model. For example, the components of EV_1 are E and V_1, both of which are contained in the model.

$$\text{logit } P(\mathbf{X}) = \alpha + \beta E + \gamma_1 V_1^2$$
$$+ \gamma_2 V_2 + \delta_1 EV_1^2$$

HWF model?

Yes, if V_1^2 is biologically meaningful

components of EV_1^2: E and V_1^2
components of V_1^2: none

No, if V_1^2 is not meaningful separately
from V_1:

model does not contain

- V_1, component of V_1^2
- EV_1, component of EV_1^2

Why require HWF model?

Answer:

HWF?	Tests for highest-order variables?
No	dependent on coding
Yes	independent of coding

$$\text{logit } P(\mathbf{X}) = \alpha + \beta E + \gamma_1 V_1 + \gamma_2 V_2$$
$$+ \delta_1 EV_1 + \delta_2 EV_2 + \delta_3 EV_1 V_2$$

Not HWF model:

$V_1 V_2$ missing

For illustrative purposes, let us consider one other model given by logit $P(\mathbf{X})$ equals α plus βE plus $\gamma_1 V_1^2$ plus $\gamma_2 V_2$ plus the product term $\delta_1 EV_1^2$. Is this model hierarchically well formulated?

The answer here can be either *yes* or *no* depending on how the investigator wishes to treat the variable V_1^2 in the model. If V_1^2 is biologically meaningful in its own right without considering its component V_1, then the corresponding model is hierarchically well formulated because the variable EV_1^2 can be viewed as having only two components, namely, E and V_1^2, both of which are contained in the model. Also, if the variable V_1^2 is considered meaningful by itself, it can be viewed as having no lower-order components. Consequently, all lower-order components of each variable are contained in the model.

On the other hand, if the variable V_1^2 is not considered meaningful separately from its fundamental component V_1, then the model is not hierarchically well formulated. This is because, as given, the model does not contain V_1, which is a lower-order component of V_1^2 and EV_1^2, and also does not contain the variable EV_1 which is a lower-order component of EV_1^2.

Now that we have defined and illustrated an HWF model, we discuss why such a model structure is required. The reason is that if the model is not HWF, then tests about variables in the model—in particular, the highest-order terms—may give varying results depending on the coding of variables in the model. Such tests should be **independent of the coding** of the variables in the model, and they are if the model is hierarchically well formulated.

To illustrate this point, we return to the first example considered above, where the model is given by logit $P(\mathbf{X})$ equals α plus βE plus $\gamma_1 V_1$ plus $\gamma_2 V_2$ plus the product terms $\delta_1 EV_1$ plus $\delta_2 EV_2$ plus $\delta_3 EV_1 V_2$. This model is not hierarchically well formulated because it is missing the term $V_1 V_2$. The highest-order term in this model is the three-factor product term $EV_1 V_2$.

EXAMPLE (continued)

E dichotomous:

Then if *not* HWF model,
testing for EV_1V_2 may depend on
whether E is coded as

$E = (0, 1)$, e.g., significant

or

$E = (-1, 1)$, e.g., not significant

or

other coding

EXAMPLE

HWF model:
$$\text{logit P}\big(\mathbf{X}\big) = \alpha + \beta E + \gamma_1 V_1 + \gamma_2 V_2 + \delta_3 V_1 V_2$$
$$+ \delta_1 EV_1 + \delta_2 EV_2 + \delta_3 EV_1 V_2$$

Testing for EV_1V_2 is **independent of coding** of E: $(0, 1)$, $(-1, 1)$, or other.

HWF model: Tests for *lower*-order terms depend on coding

EXAMPLE

HWF model:
$$\text{logit P}\big(\mathbf{X}\big) = \alpha + \beta E + \gamma_1 V_1 + \gamma_2 V_2 + \gamma_3 V_1 V_2$$
$$+ \delta_1 EV_1 + \delta_2 EV_2 + \delta_3 EV_1 V_2$$

EV_1V_2: **not dependent** on coding

EV_1 or EV_2: **dependent** on coding

Require
- HWF model
- no test for lower-order components of significant higher-order terms

Suppose that the exposure variable E in this model is a dichotomous variable. Then, because the model is not HWF, a test of hypothesis for the significance of the highest-order term, EV_1V_2, may give different results depending on whether E is coded as $(0, 1)$ or $(-1, 1)$ or any other coding scheme.

In particular, it is possible that a test for EV_1V_2 may be highly significant if E is coded as $(0, 1)$, but be non-significant if E is coded as $(-1, 1)$. Such a possibility should be avoided because the coding of a variable is simply a way to indicate categories of the variable and, therefore, should not have an effect on the results of data analysis.

In contrast, suppose we consider the HWF model obtained by adding the V_1V_2 term to the previous model. For this model, a test for EV_1V_2 will give exactly the same result whether E is coded using $(0, 1)$, $(-1, 1)$, or any other coding. In other words, such a test is independent of the coding used.

We will shortly see that even if the model is hierarchically well formulated, then tests about lower-order terms in the model may still depend on the coding.

For example, even though, in the HWF model being considered here, a test for EV_1V_2 is not dependent on the coding, a test for EV_1 or EV_2—which are lower-order terms—may still be dependent on the coding.

What this means is that in addition to requiring that the model be HWF, we also require that no tests be allowed for lower-order components of terms like EV_1V_2 already found to be significant. We will return to this point later when we describe the hierarchy principle for retaining variables in the model.

VI. The Hierarchical Backward Elimination Approach

✓ Variable specification
✓ HWF model
Largest model considered
= initial (starting) model

Initial model ➡️ Final model

hierarchical
backward
elimination

```
┌─────────────────────────────┐
│        Initial Model        │
└─────────────────────────────┘
              │
              ▼
┌─────────────────────────────┐
│   Eliminate $EV_iE_j$ terms │
└─────────────────────────────┘
              │
              ▼
┌─────────────────────────────┐
│    Eliminate $EV_i$ terms   │
└─────────────────────────────┘
              │
              ▼
┌─────────────────────────────┐
│ Eliminate $V_i$ and $V_iV_j$ terms │
└─────────────────────────────┘
```

EV_i and EV_j (interactions):
  use statistical testing

V_i and V_iV_j (confounders):
  do *not* use statistical testing

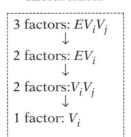

Hierarchical

3 factors: EV_iV_j
  ↓
2 factors: EV_i
  ↓
2 factors: V_iV_j
  ↓
1 factor: V_i

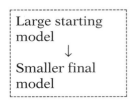

Backward

Large starting
model
  ↓
Smaller final
model

We have now completed our recommendations for variable specification as well as our requirement that the model be hierarchically well formulated. When we complete this stage, we have identified the largest possible model to be considered. This model is the initial or starting model from which we attempt to eliminate unnecessary variables.

The recommended process by which the initial model is reduced to a final model is called a **hierarchical backward elimination approach.** This approach is described by the flow diagram shown here.

In the flow diagram, we begin with the initial model determined from the variable specification stage.

If the initial model contains three-factor product terms of the form EV_iV_j, then we attempt to eliminate these terms first.

Following the three-factor product terms, we then eliminate unnecessary two-factor product terms of the form EV_i.

The last part of the strategy eliminates unnecessary V_i and V_iV_j terms.

As described in later sections, the EV_iV_j and EV_i product terms can be eliminated using appropriate statistical testing methods.

However, decisions about the V_i and V_iV_j terms, which are potential confounders, should not involve statistical testing.

The **strategy** described by this flow diagram is called **hierarchical backward** because we are working backward from our largest starting model to a smaller final and we are treating variables of different orders at different steps. That is, there is a hierarchy of variable types, with three-factor interaction terms considered first, followed by two-factor interaction terms, followed by two-factor, and then one-factor confounding terms.

VII. The Hierarchy Principle for Retaining Variables

Hierarchical Backward Elimination

retain terms   drop terms
↓
Hierarchy Principle

(à la Bishop, Fienberg, and Holland)

retain lower-order components

As we go through the hierarchical backward elimination process, some terms are retained and some terms are dropped at each stage. For those terms that are retained at a given stage, there is a rule for identifying lower-order components that must also be retained in any further models.

This rule is called the **Hierarchy Principle.** An analogous principle of the same name has been described by Bishop, Fienberg, and Holland in their textbook *Discrete Multivariate Analysis: Theory and Practice*, MIT Press, 1975.

EXAMPLE

Initial model: EV_iV_j terms

Suppose: EV_2V_5 significant

Hiearchy Principle: all lower-order components of EV_2V_5 retained

i.e., E, V_2, V_5, EV_2, EV_5, and V_2V_5 cannot be eliminated

Note: Initial model must contain V_2V_5 to be HWF

To illustrate the Hierarchy Principle, suppose the initial model contains three-factor products of the form EV_iV_j. Suppose, further, that the term EV_2V_5 is found to be significant during the stage which considers the elimination of unimportant EV_iV_j terms. Then, the Hierarchy Principle requires that all lower-order components of the EV_2V_5 term must be retained in all further models considered in the analysis.

The lower-order components of EV_2V_5 are the variables E, V_2, V_5, EV_2, EV_5, and V_2V_5. Because of the Hierarchy Principle, if the term EV_2V_5 is retained, then each of the above component terms cannot be eliminated from all further models considered in the backward elimination process. Note the initial model has to contain each of these terms, including V_2V_5, to ensure that the model is hierarchically well formulated.

Hierarchy Principle

If product variable retained, then *all* lower-order components must be retained

In general, the Hierarchy Principle states that if a product variable is retained in the model, then all lower-order components of that variable must be retained in the model.

EXAMPLE

EV_2 and EV_4 retained:
Then
E, V_2 and V_4 also retained

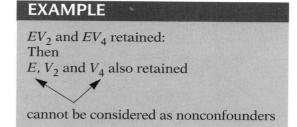

cannot be considered as nonconfounders

As another example, if the variables EV_2 and EV_4 are to be retained in the model, then the following lower-order components must also be retained in all further models considered: E, V_2, and V_4. Thus, we are not allowed to consider dropping V_2 and V_4 as possible nonconfounders because these variables must stay in the model regardless.

Hiearchy Principle rationale

- tests for lower-order components depend on coding
- tests should be independent of coding
- therefore, no tests allowed for lower-order components

EXAMPLE

Suppose EV_2V_5 significant: then the test for EV_2 depends on coding of E, e.g., $(0, 1)$ or $(-1, 1)$

HWF model:

tests for *highest-order* terms *independent* of coding

but

tests for *lower-order* terms *dependent* on coding

EXAMPLE

HWF: EV_iV_j highest-order terms
Then tests for
   EV_iV_j **independent** of coding **but**
tests for
   EV_i or V_j **dependent** on coding

EXAMPLE

HWF: EV_i highest-order terms
Then tests for
   EV_i **independent** of coding **but** tests for
   V_i **dependent** on coding

Hierarchy Principle

- ensures that the model is HWF

e.g., EV_iV_j is significant $\Rightarrow$ retain lower-order components or else model is not HWF

The **rationale for the Hierarchy Principle** is similar to the rationale for requiring that the model be HWF. That is, tests about lower-order components of variables retained in the model can give different conclusions depending on the coding of the variables tested. Such tests should be independent of the coding to be valid. Therefore, no such tests are appropriate for lower-order components.

For example, if the term EV_2V_5 is significant, then a test for the significance of EV_2 may give different results depending on whether E is coded as $(0, 1)$ or $(-1, 1)$.

Note that if a model is HWF, then tests for the highest-order terms in the model are always independent of the coding of the variables in the model. However, tests for lower-order components of higher-order terms are still dependent on coding.

For example, if the highest-order terms in an HWF model are of the form EV_iV_j, then tests for all such terms are not dependent on the coding of any of the variables in the model. However, tests for terms of the form EV_i or V_i are dependent on the coding and, therefore, should not be carried out as long as the corresponding higher-order terms remain in the model.

If the highest-order terms of a hierarchically well-formulated model are of the form EV_i, then tests for EV_i terms are independent of coding, but tests for V_i terms are dependent on coding of the V's and should not be carried out. Note that because the V's are potential confounders, tests for V's are not allowed anyhow.

Note also, regarding the Hierarchy Principle, that any lower-order component of a significant higher-order term must remain in the model or else the model will no longer be HWF. Thus, to ensure that our model is HWF as we proceed through our strategy, we cannot eliminate lower-order components unless we have eliminated corresponding higher-order terms.

VIII. An Example

Cardiovascular Disease Study
9-year follow-up Evans County, GA
$n = 609$ white males

The variables:

CAT, AGE, CHL, SMK, ECG, HPT

at start
CHD = outcome

CAT: (0, 1) exposure
AGE, CHL: continuous ⎱ control
SMK, ECG, HPT: (0, 1) ⎰ variables

$E = \text{CAT}$ [?] ⟩ $D = \text{CHD}$

controlling for

AGE, CHL, SMK, ECG, HPT

C's

Variable specification stage:
V's: potential confounders in initial model

Here, V's = C's:

$V_1 = \text{AGE}, V_2 = \text{CHL}, V_3 = \text{SMK},$
$V_4 = \text{ECG}, V_5 = \text{HPT}$

Other possible V's:
$V_6 = \text{AGE} \times \text{CHL}$
$V_7 = \text{AGE} \times \text{SMK}$
$V_8 = \text{AGE}^2$
$V_9 = \text{CHL}^2$

We review the guidelines recommended to this point through an example. We consider a cardiovascular disease study involving the 9-year follow-up of persons from Evans County, Georgia. We focus on data involving 609 white males on which we have measured 6 variables at the start of the study. These are catecholamine level (CAT), AGE, cholesterol level (CHL), smoking status (SMK), electrocardiogram abnormality status (ECG), and hypertension status (HPT). The outcome variable is coronary heart disease status (CHD).

In this study, the exposure variable is CAT, which is 1 if high and 0 if low. The other five variables are control variables, so that these may be considered as confounders and/or effect modifiers. AGE and CHL are treated continuously, whereas SMK, ECG, and HPT, are (0, 1) variables.

The question of interest is to describe the relationship between E (CAT) and D (CHD), controlling for the possible confounding and effect modifying effects of AGE, CHL, SMK, ECG, and HPT. These latter five variables are the C's that we have specified at the start of our modeling strategy.

To follow our strategy for dealing with this data set, we now carry out variable specification in order to define the initial model to be considered. We begin by specifying the V variables, which represent the potential confounders in the initial model.

In choosing the V's, we follow our earlier recommendation to let the V's be the same as the C's. Thus, we will let $V_1 = \text{AGE}, V_2 = \text{CHL}, V_3 = \text{SMK}, V_4 = \text{ECG},$ and $V_5 = \text{HPT}$.

We could have chosen other V's in addition to the five C's. For example, we could have considered V's which are products of two C's, such as V_6 equals AGE $\times$ CHL or V_7 equals AGE $\times$ SMK. We could also have considered V's which are squared C's, such as V_8 equals AGE^2 or V_9 equals CHL^2.

EXAMPLE (continued)

Restriction of V's to C's because
- large number of C's
- additional V's difficult to interpret
- additional V's may lead to collinearity

Choice of W's:
(go into model as EW)
  W's = C's:
  W_1 = AGE, W_2 = CHL, W_3 = SMK, W_4 = ECG, W_5 = HPT

Other possible W's:
  W_6 = AGE × CHL
(If W_6 is in model, then V_6 = AGE × CHL also in HWF model.)

Alternative choice of W's:
Subset of C's, e.g.,

  AGE ⇒ CAT × AGE in model

  ECG ⇒ CAT × ECG in model

Rationale for W's = C's:
- allow possible interaction
- minimize collinearity

Initial E, V, W model

$$\text{logit P}(\mathbf{X}) = \alpha + \beta\text{CAT} + \sum_{i=1}^{5} \gamma_i V_i + \text{CAT} \sum_{j=1}^{5} \delta_j W_j$$

where V_i's = C's = W_j's

However, we have restricted the V's to the C's themselves primarily because there are a moderately large number of C's being considered, and any further addition of V's is likely to make the model difficult to interpret as well as difficult to fit because of likely collinearity problems.

We next choose the W's, which are the variables that go into the initial model as product terms with E(CAT). These W's are the potential effect modifiers to be considered. The W's that we choose are the C's themselves, which are also the V's. That is, W_1 through W_5 equals AGE, CHL, SMK, ECG, and HPT, respectively.

We could have considered other choices for the W's. For instance, we could have added two-way products of the form W_6 equals AGE × CHL. However, if we added such a term, we would have to add a corresponding two-way product term as a V variable, that is, V_6 equals AGE × CHL, to make our model hierarchically well formulated. This is because AGE × CHL is a lower-order component of CAT × AGE × CHL, which is EW_6.

We could also have considered for our set of W's some subset of the five C's, rather than all five C's. For instance, we might have chosen the W's to be AGE and ECG, so that the corresponding product terms in the model are CAT × AGE and CAT × ECG only.

Nevertheless, we have chosen the W's to be all five C's so as to consider the possibility of interaction from any of the five C's, yet to keep the model relatively small to minimize potential collinearity problems.

Thus, at the end of the variable specification stage, we have chosen as our initial model, the E, V, W model shown here. This model is written in logit form as logit P($\mathbf{X}$) equals a constant term plus terms involving the main effects of the five control variables plus terms involving the interaction of each control variable with the exposure variable CAT.

EXAMPLE (continued)

HWF model?

    i.e., given variable, are lower-order components in model?

    e.g., CAT × AGE
        ⇓

CAT and AGE both in model as main effects

HWF model?   **YES**

If CAT × ECG × SMK in model, then **not** HWF model
because
ECG × SMK not in model

Next

    Hierarchical Backward Elimination Procedure

First, eliminate *EW* terms

Then, eliminate *V* terms

Interaction assessment
  and
confounding assessments (details in Chapter 7)

Results of Interaction Stage
CAT × CHL and CAT × HPT
are the only two interaction terms to remain in the model
Model contains
  CAT, AGE, CHL, SMK, ECG, HPT,
            V's
  CAT × CHL and CAT × HPT

According to our strategy, it is necessary that our initial model, or any subsequently determined reduced model, be hierarchically well formulated. To check this, we assess whether all lower-order components of any variable in the model are also in the model.

For example, the lower-order components of a product variable like CAT × AGE are CAT and AGE, and both these terms are in the model as main effects. If we identify the lower-order components of any other variable, we can see that the model we are considering is truly hierarchically well formulated.

Note that if we add to the above model the three-way product term CAT × ECG × SMK, the resulting model is not hierarchically well formulated. This is because the term ECG × SMK has not been specified as one of the *V* variables in the model.

At this point in our model strategy, we are ready to consider simplifying our model by eliminating unnecessary interaction and/or confounding terms. We do this using a hierarchical backward elimination procedure which considers eliminating the highest-order terms first, then the next highest-order terms, and so on.

Because the highest-order terms in our initial model are two-way products of the form *EW*, we first consider eliminating some of these interaction terms. We then consider eliminating the *V* terms, which are the potential confounders.

Here, we summarize the results of the interaction assessment and confounding assessment stages and then return to provide more details of this example in Chapter 7.

The results of the interaction stage allow us to eliminate three interaction terms, leaving in the model the two product terms CAT × CHL and CAT × HPT.

Thus, at the end of interaction assessment, our remaining model contains our exposure variable CAT, the five *V*'s namely, AGE, CHL, SMK, ECG, and HPT plus two product terms CAT × CHL and CAT × HPT.

EXAMPLE (continued)

All five V's in model so far

Hierarchy Principle

   identify V's that **cannot** be eliminated
        EV_i significant
              $\Downarrow$
     E and V_i must remain

CAT $\times$ CHL $\Rightarrow$ CAT and CHL remain
CAT $\times$ HPT $\Rightarrow$ CAT and HPT remain

Thus,
   CAT (exposure) remains
plus
   CHL and HPT remain

AGE, SMK, ECG
   eligible for elimination

Results (details in Chapter 7):

Cannot remove AGE, SMK, ECG
   (decisions too subjective)

Final model variables:
   CAT, AGE, CHL, SMK, ECG, HPT,
   CAT $\times$ CHL, and CAT $\times$ HPT

The reason why the model contains all five V's at this point is because we have not yet done any analysis to evaluate which of the V's can be eliminated from the model.

However, because we have found two significant interaction terms, we need to use the Hierarchy Principle to identify certain V's that cannot be eliminated from any further models considered.

The hierarchy principle says that all lower-order components of significant product terms must remain in all further models.

In our example, the lower-order components of CAT $\times$ CHL are CAT and CHL, and the lower-order components of CAT $\times$ HPT are CAT and HPT. Now the CAT variable is our exposure variable, so we will leave CAT in all further models regardless of the hierarchy principle. In addition, we see that CHL and HPT must remain in all further models considered.

This leaves the V variables AGE, SMK, and ECG as still being eligible for elimination at the confounding stage of the strategy.

As we show in Chapter 7, we will not find sufficient reason to remove any of the above three variables as nonconfounders. In particular, we will show that decisions about confounding for this example are too subjective to allow us to drop any of the three V terms eligible for elimination.

Thus, as a result of our modeling strategy, the final model obtained contains the variables CAT, AGE, CHL, SMK, ECG, and HPT as main effect variables, and it contains the two product terms CAT $\times$ CHL and CAT $\times$ HPT.

At this stage one still hasn't ran any modelling programs (not yet sought α_3, β_{ij})

EXAMPLE (continued)

Printout

Variable	Coefficient	S.E.	Chi sq	P
Intercept	−4.0474	1.2549	10.40	0.0013
CAT	−12.6809	3.1042	16.69	0.0000
AGE	0.0349	0.0161	4.69	0.0303
CHL	−0.0055	0.0042	1.70	0.1917
ECG	0.3665	0.3278	1.25	0.2635
SMK	0.7735	0.3272	5.59	0.0181
HPT	1.0468	0.3316	9.96	0.0016
CH	−2.3299	0.7422	9.85	0.0017
CC	0.0691	0.0143	23.18	0.0000

V's { AGE, CHL, ECG, SMK, HPT }

interaction { CH, CC }

interaction

CH = CAT × HPT and
CC = CAT × CHL

$$\widehat{ROR} = \exp(-12.6809 + 0.0691 CHL - 2.3299 HPT)$$

Details in Chapter 7.

The computer results for this final model are shown here. This includes the estimated regression coefficients, corresponding standard errors, and Wald test information. The variables CAT × HPT and CAT × CHL are denoted in the printout as CH and CC, respectively.

Also provided here is the formula for the estimated adjusted odds ratio for the CAT, CHD relationship. Using this formula, one can compute point estimates of the odds ratio for different specifications of the effect modifiers CHL and HPT. Further details of these results, including confidence intervals, will be provided in Chapter 7.

SUMMARY

Three stages

(1) variable specification
(2) interaction
(3) confounding/precision

Initial model: HWF model

Hierarchical backward elimination procedure
(test for interaction, but do not test for confounding)

Hierarchy Principle
significant product term
⇓
retain lower-order components

As a summary of this presentation, we have recommended a modeling strategy with three stages: (1) **variable specification,** (2) **interaction assessment,** and (3) **confounding assessment** followed by consideration of **precision.**

The initial model has to be **hierarchically well formulated** (HWF). This means that the model must contain all lower-order components of any term in the model.

Given an initial model, the recommended strategy involves a **hierarchical backward elimination procedure** for removing variables. In carrying out this strategy, statistical testing is allowed for interaction terms, but not for confounding terms.

When assessing interaction terms, the **Hierarchy Principle** needs to be applied for any product term found significant. This principle requires all lower-order components of significant product terms to remain in all further models considered.

Chapters up to this point

This presentation is now complete. We suggest that the reader review the presentation through the detailed outline on the following pages. Then, work through the practice exercises and then the test.

The next chapter is entitled: "Modeling Strategy for Assessing Interaction and Confounding." This continues the strategy described here by providing a detailed description of the interaction and confounding assessment stages of our strategy.

**Detailed
Outline**

I. **Overview** (page 164)

Focus:

- guidelines for "best" model
- 3-stage strategy
- valid estimate of E–D relationship

II. **Rationale for a modeling strategy** (pages 164–165)

A. Insufficient explanation provided about strategy in published research; typically only final results are provided.

B. Too many ad hoc strategies in practice; need for some guidelines.

C. Need to consider a general strategy that applies to different kinds of modeling procedures.

D. Goal of strategy in etiologic research is to get a valid estimate of E–D relationship; this contrasts with goal of obtaining good prediction, which is built into computer packages for different kinds of models.

III. **Overview of recommended strategy** (pages 165–169)

A. Three stages: variable specification, interaction assessment, and confounding assessment followed by considerations of precision.

B. Reason why interaction stage precedes confounding stage: confounding is irrelevant in the presence of strong interaction.

C. Reason why confounding stage considers precision after confounding is assessed: validity takes precedence over precision.

D. Statistical concerns needing attention but beyond scope of this presentation: collinearity, controlling the significance level, and influential observations.

E. The model must be hierarchically well formulated.

F. The strategy is a hierarchical backward elimination strategy that considers the roles that different variables play in the model and cannot be directly carried out using standard computer algorithms.

G. Confounding is not assessed by statistical testing.

H. If interaction is present, confounding assessment is difficult in practice.

IV. **Variable specification stage** (pages 169–171)

A. Start with D, E, and $C_1, C_2, \ldots, C_p$.

B. Choose V's from C's based on prior research or theory and considering potential statistical problems, e.g., collinearity; simplest choice is to let V's be C's themselves.

C. Choose W's from C's to be either V's or product of two V's; usually recommend W's to be C's themselves or some subset of C's.

V. Hierarchically well-formulated (HWF) models (pages 171–173)

    A. Definition: given any variable in the model, all lower-order components must also be in the model.

    B. Examples of models that are and are not hierarchically well formulated.

    C. Rationale: If model is not hierarchically well formulated, then tests for significance of the highest-order variables in the model may change with the coding of the variables tested; such tests should be independent of coding.

VI. The hierarchical backward elimination approach (page 174)

    A. Flow diagram representation.

    B. Flow description: evaluate EV_iV_j terms first, then EV_i terms, then V_i terms last.

    C. Use statistical testing for interaction terms, but decisions about V_i terms should not involve testing.

VII. The Hierarchy Principle for retaining variables (pages 175–176)

    A. Definition: If a variable is to be retained in the model, then all lower-order components of that variable are to be retained in the model forever.

    B. Example.

    C. Rationale: Tests about lower-order components can give different conclusions depending on the coding of variables tested; such tests should be independent of coding to be valid; therefore, no such tests are appropriate.

    D. Example.

VIII. An example (pages 177–181)

    A. Evans County CHD data description.

    B. Variable specification stage.

    C. Final results.

Practice Exercises

A prevalence study of predictors of surgical wound infection in 265 hospitals throughout Australia collected data on 12,742 surgical patients (McLaws et al., *Med. J. of Australia*, 1988). For each patient, the following independent variables were determined: type of hospital (public or private), size of hospital (large or small), degree of contamination of surgical site (clean or contaminated), and age and sex of the patient. A logistic model was fitted to these data to predict whether or not the patient developed a surgical wound infection during hospitalization. The abbreviated variable names and the manner in which the variables were coded in the model are described as follows:

Variable	Abbreviation	Coding
Type of hospital	HT	1 = public, 0 = private
Size of hospital	HS	1 = large, 0 = small
Degree of contamination	CT	1 = contaminated, 0 = clean
Age	AGE	Continuous
Sex	SEX	1 = female, 0 = male

In the questions that follow, we assume that type of hospital (HT) is considered the exposure variable, and the other four variables are risk factors for surgical wound infection to be considered for control.

1. In defining an E, V, W model to describe the effect of HT on the development of surgical wound infection, describe how you would determine the V variables to go into the model. (In answering this question, you need to specify the criteria for choosing the V variables, rather than the specific variables themselves.)

2. In defining an E, V, W model to describe the effect of HT on the development of surgical wound infection, describe how you would determine the W variables to go into the model. (In answering this question, you need to specify the criteria for choosing the W variables, rather than the specifying the actual variables.)

3. State the logit form of a hierarchically well-formulated E, V, W model for the above situation in which the V's and the W's are the C's themselves. Why is this model hierarchically well formulated?

4. Suppose the product term HT $\times$ AGE $\times$ SEX is added to the model described in Exercise 3. Is this new model still hierarchically well formulated? If so, state why; if not, state why not.

5. Suppose for the model described in Exercise 4, that a Wald test is carried out for the significance of the three-factor product term HT $\times$ AGE $\times$ SEX. Explain what is meant by the statement that the test result depends on the coding of the variable HT. Should such a test be carried out? Explain briefly.

6. Suppose for the model described in Exercise 3 that a Wald test is carried out for the significance of the two-factor product term HT $\times$ AGE. Is this test dependent on coding? Explain briefly.

7. Suppose for the model described in Exercise 3 that a Wald test is carried out for the significance of the main effect term AGE. Why is this test inappropriate here?

8. Using the model of Exercise 3, describe briefly the hierarchical backward elimination procedure for determining the best model.

9. Suppose the interaction assessment stage for the model of Example 3 finds the following two-factor product terms to be significant: HT $\times$ CT

and HT $\times$ SEX; the other two-factor product terms are not significant and are removed from the model. Using the Hierarchy Principle, what variables must be retained in all further models considered. Can these (latter) variables be tested for significance? Explain briefly.

10. Based on the results in Exercise 9, state the (reduced) model that is left at the end of the interaction assessment stage.

Test

True or False? (Circle T or F)

T   F   1. The three stages of the modeling strategy described in this chapter are interaction assessment, confounding assessment, and precision assessment.

T   F   2. The assessment of interaction should precede the assessment of confounding.

T   F   3. The assessment of interaction may involve statistical testing.

T   F   4. The assessment of confounding may involve statistical testing.

T   F   5. Getting a precise estimate takes precedence over getting an unbiased answer.

T   F   6. During variable specification, the potential confounders should be chosen based on analysis of the data under study.

T   F   7. During variable specification, the potential effect modifiers should be chosen by considering prior research or theory about the risk factors measured in the study.

T   F   8. During variable specification, the potential effect modifiers should be chosen by considering possible statistical problems that may result from the analysis.

T   F   9. A model containing the variables E, A, B, C, A^2, $A \times B$, $E \times A$, $E \times A^2$, $E \times A \times B$, and $E \times C$ is hierarchically well formulated.

T   F   10. If the variables $E \times A^2$ and $E \times A \times B$ are found to be significant during interaction assessment, then a *complete* list of all components of these variables that must remain in any further models considered consists of E, A, B, $E \times A$, $E \times B$, and A^2.

The following questions consider the use of logistic regression on data obtained from a matched case-control study of cervical cancer in 313 women from Sydney, Australia (Brock et al., *J. Nat. Cancer Inst.*, 1988). The outcome variable is cervical cancer status (1 = present, 0 = absent). The matching variables are age and socioeconomic status. Additional independent variables not matched on are smoking status, number of lifetime sexual partners, and age at first sexual intercourse. The independent variables are listed below together with their computer abbreviation and coding scheme.

Variable	Abbreviation	Coding
Smoking status	SMK	1 = ever, 0 = never
Number of sexual partners	NS	1 = 4+, 0 = 0–3
Age at first intercourse	AS	1 = 20+, 0 = <19
Age of subject	AGE	Category matched
Socioeconomic status	SES	Category matched

11. Consider the following E, V, W model that considers the effect of smoking, as the exposure variable, on cervical cancer status, controlling for the effects of the other four independent variables listed:

$$\text{logit P}(\mathbf{X}) = \alpha + \beta\text{SMK} + \sum \gamma_i^* V_i^* + \gamma_1 \text{NS} + \gamma_2 \text{AS} + \gamma_3 \text{NS} \times \text{AS}$$
$$+ \delta_1 \text{SMK} \times \text{NS} + \delta_2 \text{SMK} \times \text{AS} + \delta_3 \text{SMK} \times \text{NS} \times \text{AS},$$

where the V_i^* are dummy variables indicating matching strata and the γ_i^* are the coefficients of the V_i^* variables. Is this model hierarchically well formulated? If so, explain why; if not, explain why not.

12. For the model in Question 1, is a test for the significance of the three-factor product term SMK × NS × AS dependent on the coding of SMK? If so, explain why; if not explain, why not.

13. For the model in Question 1, is a test for the significance of the two-factor product term SMK × NS dependent on the coding of SMK? If so, explain why; if not, explain why not.

14. For the model in Question 1, briefly describe a hierarchical backward elimination procedure for obtaining a best model.

15. Suppose that the three-factor product term SMK × NS × AS is found significant during the interaction assessment stage of the analysis. Then, using the Hierarchy Principle, what other *interaction* terms must remain in any further model considered? Also, using the Hierarchy Principle, what *potential confounders* must remain in any further models considered?

16. Assuming the scenario described in Question 15 (i.e., SMK × NS × AS is significant), what (reduced) model remains after the interaction assessment stage of the model? Are there any potential confounders that are still eligible to be dropped from the model. If so, which ones? If not, why not?

Answers to Practice Exercises

1. The V variables should include the C variables HS, CT, AGE, and SEX and any functions of these variables that have some justification based on previous research or theory about risk factors for surgical wound infection. The simplest choice is to choose the V's to be the C's themselves, so that at least every variable already identified as a risk factor is controlled in the simplest way possible.

2. The W variables should include some subset of the V's, or possibly all the V's, plus those functions of the V's that have some support from prior research or theory about effect modifiers in studies of surgical wound infection. Also, consideration should be given, when choosing the W's, of possible statistical problems, e.g., collinearity, that may arise if the size of the model becomes quite large and the variables chosen are higher-order product terms. Such statistical problems may be avoided if the W's chosen do not involve very high-order product terms and if the number of W's chosen is small. A safe choice is to choose the W's to be the V's themselves or a subset of the V's.

3. logit $P(\mathbf{X}) = \alpha + \beta HT + \gamma_1 HS + \gamma_2 CT + \gamma_3 AGE + \gamma_4 SEX + \delta_1 HT \times HS + \delta_2 HT \times CT + \delta_3 HT \times AGE + \delta_4 HT \times SEX.$
    This model is HWF because given any interaction term in the model, both of its components are also in the model (as main effects).

4. If $HT \times AGE \times SEX$ is added to the model, the new model will *not* be hierarchically well formulated because the lower-order component $AGE \times SEX$ is not contained in the original nor new model.

5. A test for $HT \times AGE \times SEX$ in the above model is dependent on coding in the sense that different test results (e.g., rejection versus nonrejection of the null hypothesis) may be obtained depending on whether HT is coded as $(0, 1)$ or $(-1, 1)$ or some other coding. Such a test should not be carried out because any test of interest should be independent of coding, reflecting whatever the real effect of the variable is.

6. A test for $HT \times AGE$ in the model of Exercise 3 is independent of coding because the model is hierarchically well formulated and the $HT \times AGE$ term is a variable of highest order in the model. (Tests for lower-order terms like HT or HS are dependent on the coding even though the model in Exercise 3 is hierarchically well formulated.)

7. A test for the variable AGE is inappropriate because there is a higher-order term, $HT \times AGE$, in the model, so that a test for AGE is dependent on the coding of the HT variable. Such a test is also inappropriate because AGE is a potential confounder, and confounding should not be assessed by statistical testing.

8. A hierarchical backward elimination procedure for the model in Exercise 3 would involve first assessing interaction involving the four interaction terms and then considering confounding involving the four potential confounders. The interaction assessment could be done using statistical testing, whereas the confounding assessment should not use statistical testing. When considering confounding, any V variable which is a lower-order component of a significant interaction term must remain in all further models and is not eligible for deletion as a nonconfounder. A test for any of these latter V's is inappropriate because such a test would be dependent on the coding of any variable in the model.

9. If $HT \times CT$ and $HT \times SEX$ are found significant, then the V variables CT and SEX cannot be removed from the model and must, therefore, be retained in all further models considered. The HT variable remains in all further models considered because it is the exposure variable of interest.

10. At the end of the interaction assessment stage, the remaining model is given by

    $\text{logit } P(\mathbf{X}) = \alpha + \beta HT + \gamma_1 HS + \gamma_2 CT + \gamma_3 AGE + \gamma_4 SEX + \delta_2 HT \times CT + \delta_4 HT \times SEX.$

7

Modeling Strategy for Assessing Interaction and Confounding

Introduction

This chapter continues the previous chapter (Chapter 6) that gives general guidelines for a strategy for determining a best model using a logistic regression procedure. The focus of this chapter is the interaction and confounding assessment stages of the model building strategy.

We begin by reviewing the previously recommended (Chapter 6) three-stage strategy. The initial model is required to be hierarchically well formulated. In carrying out this strategy, statistical testing is allowed for assessing interaction terms but is not allowed for assessing confounding.

For any interaction term found significant, a Hierarchy Principle is required to identify lower-order variables which must remain in all further models considered. A flow diagram is provided to describe the steps involved in interaction assessment. Methods for significance testing for interaction terms are provided.

Confounding assessment is then described, first when there is no interaction, and then when there is interaction; the latter often being difficult to accomplish in practice.

Finally, an application of the use of the entire recommended strategy is described, and a summary of the strategy is given.

Abbreviated Outline

The outline below gives the user a preview of the material to be covered in this chapter. A detailed outline for review purposes follows the presentation.

Objectives   Upon completing this chapter, the learner should be able to:

1. Describe and apply the interaction assessment stage in a particular logistic modeling situation.
2. Describe and apply the confounding assessment stage in a particular logistic modeling situation
    a. when there is no interaction;
    b. when there is interaction.

Presentation

I. Overview

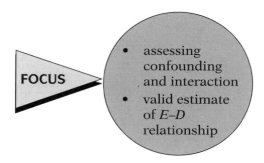

FOCUS

- assessing confounding and interaction
- valid estimate of *E–D* relationship

This presentation describes a strategy for assessing interaction and confounding when carrying out mathematical modeling using logistic regression. The goal of the strategy is to obtain a valid estimate of an exposure–disease relationship that accounts for confounding and effect modification.

Three stages
(1) variable specificaton
(2) interaction
(3) confounding/precision

In the previous presentation on modeling strategy guidelines, we recommended a modeling strategy with three stages: (1) **variable specification,** (2) **interaction assessment,** and (3) **confounding assessment** followed by consideration of **precision.**

Initial model: HWF

The initial model is required to be **hierarchically well formulated,** which we denote as HWF. This means that the initial model must contain all lower-order components of any term in the model.

EV_iV_j in initial model $\rightarrow$ $EV_i, EV_j,$ V_i, V_j, V_iV_j also in model

Thus, for example, if the model contains an interaction term of the form EV_iV_j, this will require the lower-order terms EV_i, EV_j, V_i, V_j, and V_iV_j also to be in the initial model.

Hierarchical backward elimination:

- can test for interaction, *but* not confounding
- can eliminate lower-order term if corresponding higher-order term is not significant

Given an initial model that is HWF, the recommended strategy then involves a **hierarchical backward elimination procedure** for removing variables. In carrying out this strategy, statistical testing is allowed for interaction terms but not for confounding terms. Note that although any lower-order component of a higher-order term must belong to the initial HWF model, such a component might be dropped from the model eventually if its corresponding higher-order term is found to be nonsignificant during the backward elimination process.

Hierarchy Principle:

Significant product term	$\rightarrow$	All lower-order components remain

If, however, when assessing interaction, a product term is found significant, the **Hierarchy Principle** must be applied for lower-order components. This principle requires all lower-order components of significant product terms to remain in **all** further models considered.

II. Interaction Assessment Stage

Start with HWF model

Use hierarchical backward elimination:

EV_iV_j before EV_i

Interaction stage flow:

Initial model: E, V_i, EV_i, EV_iV_j Eliminate nonsignificant EV_iV_j terms

$\downarrow$

Use Hierarchy Principle to specify for *all further models* EV_i components of significant EV_iV_j terms

$\downarrow$

Other EV_i terms: eliminate nonsignificant EV_i terms from model retaining   • significant EV_iV_j terms   • EV_i components   • V_i (or V_iV_j) terms

Statistical testing
Chunk test for entire collection of interaction terms

According to our strategy, we consider interaction after we have specified our initial model, which must be hierarchically well formulated (HWF). To address interaction, we use a hierarchical backward elimination procedure, treating higher-order terms of the form EV_iV_j prior to considering lower-order terms of the form EV_i.

A flow diagram for the interaction stage is presented here. If our initial model contains terms up to the order EV_iV_j, elimination of these latter terms is considered first. This can be achieved by statistical testing in a number of ways, which we discuss shortly.

When we have completed our assessment of EV_iV_j terms, the next step is to use the Hierarchy Principle to specify any EV_i terms which are components of significant EV_iV_j terms. Such EV_i terms are to be retained in all further models considered.

The next step is to evaluate the significance of EV_i terms other than those identified by the Hierarchy Principle. Those EV_i terms that are *nonsignificant* are eliminated from the model. For this assessment, previously significant EV_iV_j terms, their EV_i components, and all V_i terms are retained in any model considered. Note that some of the V_i terms will be of the form V_iV_j if the initial model contains EV_iV_j terms.

In carrying out statistical testing of interaction terms, we recommend that a single "chunk" test for the entire collection (or "chunk") of interaction terms of a given order be considered first.

Like Page 130

Like Page 130

EXAMPLE

EV_1V_2, EV_1V_3, EV_2V_3 in model

chunk test for $H_0 : \delta_1 = \delta_2 = \delta_3 = 0$

use LR statistic $\sim \chi_3^2$ comparing

> full model: all V_i, V_iV_j, EV_i, EV_iV_j with
> reduced model: V_i, V_iV_j, EV_i

For example, if there are a total of three EV_iV_j terms in the initial model, namely, EV_1V_2, EV_1V_3, and EV_2V_3, then the null hypothesis for this chunk test is that the coefficients of these variables, say δ_1, δ_2, and δ_3 are all equal to zero. The test procedure is a likelihood ratio (LR) test involving a chi-square statistic with three degrees of freedom which compares the full model containing all V_i, V_iV_j, EV_i, and EV_iV_j terms with a reduced model containing only V_i, V_iV_j, and EV_i terms.

Chunk Test

not significant ↙ ↘ significant

↓ ↓

eliminate all terms in chunk   retain some terms in chunk

↓

use backward elimination to eliminate terms from chunk

If the chunk test **is not significant,** then the investigator may decide to eliminate from the model all terms tested in the chunk, for example, all EV_iV_j terms. If the chunk test **is significant,** then this means that some, but not necessarily all terms in the chunk, are significant and must be retained in the model.

To determine which terms are to be retained, the investigator may carry out a backward elimination algorithm to eliminate insignificant variables from the model one at a time. Depending on the preference of the investigator, such a backward elimination procedure may be carried out without even doing the chunk test or regardless of the results of the chunk test.

EXAMPLE

HWF model:

$\boxed{EV_1V_2, EV_1V_3,}$
$V_1, V_2, V_3, V_1V_2, V_1V_3,$
EV_1, EV_2, EV_3

As an example of such a backward algorithm, suppose we again consider a hierarchically well-formulated model which contains the two EV_iV_j terms EV_1V_2 and EV_1V_3 in addition to the lower-order components V_1, V_2, V_3, V_1V_2, V_1V_3, and EV_1, EV_2, EV_3.

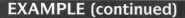

EXAMPLE (continued)

Backward approach:

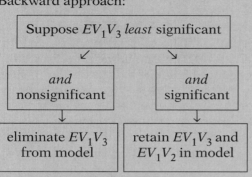

Suppose EV_1V_3 not significant:
then drop EV_1V_3 from model.
Reduced model:
EV_1V_2
$V_1, V_2, V_3, V_1V_2, V_1V_3$
EV_1, EV_2, EV_3

Suppose EV_1V_2 significant:
then EV_1V_2 retained and above
reduced model is current model

Next: eliminate EV terms

From Hierarchy Principle:
$E, V_1, V_2, EV_1, EV_2,$ and V_1V_2
retained **in all further models**

Assess other EV_i terms:
only EV_3 eligible for removal

Using the backward approach, the least significant EV_iV_j term, say EV_1V_3, is eliminated from the model first, provided it is nonsignificant, as shown on the left-hand side of the flow. If it is significant, as shown on the right-hand side of the flow, then both EV_1V_3 and EV_1V_2 must remain in the model as do all lower-order components and the modeling process is complete.

Suppose that the EV_1V_3 term is not significant. Then, this term is dropped from the model. A reduced model containing the remaining EV_1V_2 term and all lower-order components from the initial model is then fitted. The EV_1V_2 term is then dropped if nonsignificant but is retained if significant.

Suppose the EV_1V_2 term is found significant, so that as a result of backward elimination, it is the only three-factor product term retained. Then the above reduced model is our current model, from which we now work to consider eliminating EV terms.

Because our reduced model contains the significant term EV_1V_2, we must require (using the Hierarchy Principle) that the lower-order components $E, V_1, V_2,$ $EV_1, EV_2,$ and V_1V_2 are retained in all further models considered.

The next step is to assess the remaining EV_i terms. In this example, there is only one EV_i term eligible to be removed, namely EV_3, because EV_1 and EV_2 are retained from the Hierarchy Principle.

EXAMPLE (continued)

LR statistic $\sim \chi_1^2$

Full model: EV_1V_2, EV_1, EV_2, $\boxed{EV_3,}$
V_1, V_2, V_3, V_1V_2, V_1V_3

Reduced model: EV_1V_2, EV_1, EV_2,
V_1, V_2, V_3, V_1V_2, V_1V_3

Wald test: $Z = \dfrac{\hat{\delta}_{EV_3}}{S_{\hat{\delta}_{EV_3}}}$

Suppose both LR and Wald tests are nonsignificant:
  then drop EV_3 from model

Interaction stage results:

EV_1V_2, EV_1, EV_2
$\left.\begin{array}{l} V_1, V_2, V_3 \\ V_1V_2, V_1V_3 \end{array}\right\}$ confounders

All V_i (and V_iV_j) remain in model after
  interaction assessment

Most situations
use *only* EV_i
product terms
$\Downarrow$
interaction assessment
less complicated
$\Downarrow$
do not need V_iV_j terms
for HWF model.

To evaluate whether EV_3 is significant, we can perform a likelihood ratio (LR) chi-square test with one degree of freedom. For this test, the two models being compared are the full model consisting of EV_1V_2, all three EV_i terms and all V_i terms, including those of the form V_iV_j, and the reduced model which omits the EV_3 term being tested. Alternatively, a Wald test can be performed using the Z statistic equal to the coefficient of the EV_3 term divided by its standard error.

Suppose that both the above likelihood ratio and Wald tests are nonsignificant. Then we can drop the variable EV_3 from the model.

Thus, at the end of the interaction assessment stage for this example, the following terms remain in the model: EV_1V_2, EV_1, EV_2, V_1, V_2, V_3, V_1V_2, and V_1V_3.

All of the V terms, including V_1V_2 and V_1V_3, in the initial model are still in the model at this point. This is because we have been assessing interaction only, whereas the V_1V_2 and V_1V_3 terms concern confounding. Note that although the V_iV_j terms are products, they are confounding terms in this model because they do not involve the exposure variable E.

Before discussing confounding, we point out that for most situations, the highest-order interaction terms to be considered are two-factor product terms of the form EV_i. In this case, interaction assessment begins with such two-factor terms and is often much less complicated to assess than when there are terms of the form EV_iV_j.

In particular, when only two-factor interaction terms are allowed in the model, then it is *not* necessary to have two-factor confounding terms of the form V_iV_j in order for the model to be hierarchically well formulated. This makes the assessment of confounding a less complicated task than when three-factor interactions are allowed.

III. Confounding and Precision Assessment When No Interaction

Confounding:

  no statistical testing
  (validity issue)

The final stage of our strategy concerns the assessment of confounding followed by consideration of precision. We have previously pointed out that this stage, in contrast to the interaction assessment stage, is carried out without the use of statistical testing. This is because confounding is a validity issue and, consequently, does not concern random error issues which characterize statistical testing.

Confounding before Precision
      ↓   ↓
 ✓gives correct   gives narrow
   answer   confidence
                     interval

We have also pointed out that controlling for confounding takes precedence over achieving precision because the primary goal of the analysis is to obtain the correct estimate rather than a narrow confidence interval around the wrong estimate.

No interaction model:

$$\text{logit P}(\mathbf{X}) = \alpha + \beta E + \sum \gamma_i V_i$$

(no terms of form EW)

In this section, we focus on the assessment of confounding when the model contains no interaction terms. The model in this case contains only E and V terms but does not contain product terms of the form E times W.

Interaction present?	Confounding assessment?
No	Straightforward
Yes	Difficult

The assessment of confounding is relatively straightforward when no interaction terms are present in one's model. In contrast, as we shall describe in the next section, it becomes difficult to assess confounding when interaction is present.

EXAMPLE

Initial model
$$\text{logit P}(\mathbf{X}) = \alpha + \beta E + \gamma_1 V_1 + \cdots + \gamma_5 V_5$$

$$\widehat{\text{OR}} = e^{\hat{\beta}}$$

(a single number)
adjusts for $V_1, ..., V_5$

In considering the no interaction situation, we first consider an example involving a logistic model with a dichotomous E variable and five V variables, namely, V_1 through V_5.

For this model, the estimated odds ratio that describes the exposure–disease relationship is given by the expression e to the β "hat," where β "hat" is the estimated coefficient of the E variable. Because the model contains no interaction terms, this odds ratio estimate is a single number which represents an adjusted estimate which controls for all five V variables.

EXAMPLE (continued)

Gold standard estimate:
    controls for all potential confounders
    (i.e., all five V's)

We refer to this estimate as the **gold standard estimate** of effect because we consider it the best estimate we can obtain which controls for **all** the potential confounders, namely, the five V's, in our model.

Other OR estimates:
    drop some V's
    e.g., drop V_3, V_4, V_5
Reduced model
$$\text{logit } P(\mathbf{X}) = \alpha + \beta E + \gamma_1 V_1 + \gamma_2 V_2$$
$$\widehat{OR} = e^{\hat{\beta}}$$
controls for V_1 and V_2 only

We can nevertheless obtain other estimated odds ratios by dropping some of the V's from the model. For example, we can drop V_3, V_4, and V_5 from the model and then fit a model containing E, V_1 and V_2. The estimated odds ratio for this "reduced" model is also given by the expression e to the β "hat," where β "hat" is the coefficient of E in the reduced model. This estimate controls for only V_1 and V_2 rather than all five V's.

reduced model $\neq$ gold standard model
    $\boxed{\text{correct answer}}$
$$\widehat{OR} \text{ (reduced) } \overset{?}{=} \widehat{OR} \text{ (gold standard)}$$

If different, then reduced model *does not* control for confounding

Because the reduced model is different from the gold standard model, the estimated odds ratio obtained for the reduced model may be meaningful different from the gold standard. If so, then we say that the reduced model does not control for confounding because it does not give us the correct answer (i.e., gold standard).

Suppose:
    Gold standard (all five V's)
        $\widehat{OR} = 2.5$
    reduced model (V_1 and V_2)
    $\uparrow$ $\widehat{OR} = 5.2$
does not control   meaningfully
for confounding   different

For example, suppose that the gold standard odds ratio controlling for all five V's is 2.5, whereas the odds ratio obtained when controlling for only V_1 and V_2 is 5.2. Then, because these are meaningfully different odds ratios, we cannot use the reduced model containing V_1 and V_2 because the reduced model does not properly control for confounding.

$$\widehat{OR}\left(\begin{array}{c}\text{some other}\\\text{subset of }V\text{'s}\end{array}\right) \overset{?}{=} \widehat{OR}\left(\begin{array}{c}\text{gold}\\\text{standard}\end{array}\right)$$

If equal, then subset controls confounding

Now although use of only V_1 and V_2 may not control for confounding, it is possible that some other subset of the V's may control for confounding by giving essentially the same estimated odds ratio as the gold standard.

$\widehat{OR}$ (V_3 alone) = 2.7
$\widehat{OR}$ (V_4 and V_5) = 2.3
$\widehat{OR}$ (gold standard) = 2.5

All three estimates are "essentially" the same as the gold standard

For example, perhaps when controlling for V_3 alone, the estimated odds ratio is 2.7 and when controlling for V_4 and V_5, the estimated odds ratio is 2.3. The use of either of these subsets controls for confounding because they give essentially the same answer as the 2.5 obtained for the gold standard.

In general, when no interaction, assess confounding by

- monitoring changes in effect measure for subsets of V's, i.e., monitor changes in
$$\widehat{OR} = e^{\hat{\beta}}$$
- identify subsets of V's giving approximately same $\widehat{OR}$ as gold standard

In general, regardless of the number of V's in one's model, the method for assessing confounding when there is no interaction is to monitor changes in the effect measure corresponding to different subsets of potential confounders in the model. That is, we must see to what extent the estimated odds ratio given by e to the β "hat" for a given subset is different from the gold standard odds ratio.

More specifically, to assess confounding, we need to identify subsets of the V's that give approximately the same odds ratio as the gold standard. Each of these subsets controls for confounding.

If $\widehat{OR}$ (subset of V's) $= \widehat{OR}$ (gold standard), then
- which subset to use?
- why not use gold standard?

If we find one or more subset of the V's which give us the same point estimate as the gold standard, how then do we decide which subset to use? Moreover, why don't we just use the gold standard?

Answer: precision

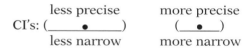

CI's:

less precise	more precise
less narrow	more narrow

The answer to both these questions involves consideration of **precision.** By precision, we refer to how narrow a confidence interval around the point estimate is. The narrower the confidence interval, the more precise the point estimate.

EXAMPLE

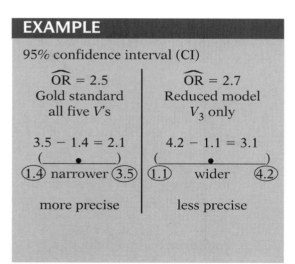

95% confidence interval (CI)

$\widehat{OR} = 2.5$	$\widehat{OR} = 2.7$
Gold standard all five V's	Reduced model V_3 only
$3.5 - 1.4 = 2.1$	$4.2 - 1.1 = 3.1$
more precise	less precise

For example, suppose the 95% confidence interval around the gold standard OR "hat" of 2.5 that controls for all five V's has limits of 1.4 and 3.5, whereas the 95% confidence interval around the OR "hat" of 2.7 that controls for V_3 only has limits of 1.1 and 4.2.

Then the gold standard OR estimate is more precise than the OR estimate that controls for V_3 only because the gold standard has the narrower confidence interval. Specifically, the narrower width is 3.5 minus 1.4, or 2.1, whereas the wider width is 4.2 minus 1.1, or 3.1.

Note that it is possible that the gold standard estimate actually may be less precise than an estimate resulting from control of a subset of V's. This will depend on the particular data set being analyzed.

Why don't we use gold standard?

Answer: Might find subset of V's which will

- gain precision (narrower CI)
- without sacrificing validity (same point estimate)

EXAMPLE		
Model	$\widehat{OR}$	CI
✓ V_4 and V_5	same (2.3)	narrower (1.9, 3.1)
Gold standard	same (2.5)	wider (1.4, 3.5)

Which subset to control?

Answer: subset with most meaningful gain in precision

Eligible subset: same point estimate as gold standard

Recommended procedure:

(1) identify eligible subsets of V's
(2) control for that subset with largest gain in precision

However, if no subset gives *better* precision, use gold standard

Scientific: Gold standard uses *all* relevant variables for control

The answer to the question, Why don't we just use the gold standard? is that we might gain a meaningful amount of precision controlling for a subset of V's, without sacrificing validity. That is, we might find a subset of V's to give essentially the same estimate as the gold standard but which also has a much narrower confidence interval.

For instance, controlling for V_4 and V_5 may obtain the same point estimate as the gold standard but a narrower confidence interval, as illustrated here. If so, we would prefer the estimate which uses V_4 and V_5 in our model to the gold standard estimate.

We also asked the question, How do we decide which subset to use for control? The answer to this is to choose that subset which gives the most meaningful gain in precision among all eligible subsets, including the gold standard.

By **eligible subset,** we mean any collection of V's that gives essentially the same point estimate as the gold standard.

Thus, we recommend the following general procedure for the confounding and precision assessment stage of our strategy:

(1) **Identify eligible subsets** of V's giving approximately the same odds ratio as the gold standard.
(2) Control for that subset which gives the largest gain in precision. However, if no subset gives meaningfully better precision than the gold standard, it is **scientifically** better to control for all V's using the gold standard.

The gold standard is **scientifically** better because persons who critically appraise the results of the study can see that when using the gold standard, all the relevant variables have been controlled for in the analysis.

EXAMPLE

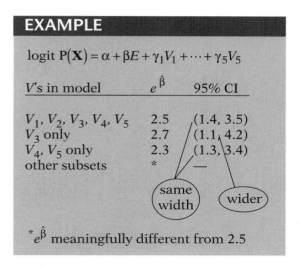

logit $P(\mathbf{X}) = \alpha + \beta E + \gamma_1 V_1 + \cdots + \gamma_5 V_5$

V's in model	$e^{\hat{\beta}}$	95% CI
V_1, V_2, V_3, V_4, V_5	2.5	(1.4, 3.5)
V_3 only	2.7	(1.1, 4.2)
V_4, V_5 only	2.3	(1.3, 3.4)
other subsets	*	—

same width   wider

$^*e^{\hat{\beta}}$ meaningfully different from 2.5

Returning to our example involving five V variables, suppose that the point estimates and confidence intervals for various subsets of V's are given as shown here. Then there are only two eligible subsets other than the gold standard—namely V_3 alone, and V_4 and V_5 together because these two subsets give the same odds ratio as the gold standard.

Considering precision, we then conclude that we should control for all five V's, that is, the gold standard, because no meaningful gain in precision is obtained from controlling for either of the two eligible subsets of V's. Note that when V_3 alone is controlled, the CI is wider than that for the gold standard. When V_4 and V_5 are controlled together, the CI is the same as the gold standard.

IV. Confounding Assessment with Interaction

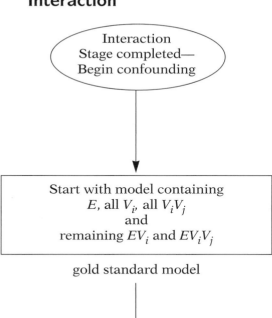

Interaction
Stage completed—
Begin confounding

Start with model containing
E, all V_i, all V_iV_j
and
remaining EV_i and EV_iV_j

gold standard model

We now consider how to assess confounding when the model contains interaction terms. A flow diagram which describes our recommended strategy for this situation is shown here. This diagram starts from the point in the strategy where interaction has already been assessed. Thus, we assume that decisions have been made about which interaction terms are significant and are to be retained in all further models considered.

In the first step of the flow diagram, we start with a model containing E and all potential confounders initially specified as V_i and V_iV_j terms plus remaining interaction terms determined from interaction assessment. This includes those EV_i and EV_iV_j terms found to be significant plus those EV_i terms that are components of significant EV_iV_j terms. Such EV_i terms must remain in all further models considered because of the Hierarchy Principle.

This model is the **gold standard model** to which all further models considered must be compared. By gold standard, we mean that the odds ratio for this model controls for all potential confounders in our initial model, that is, all the V_i's and V_iV_j's.

Apply Hierarchy Principle to identify V_i, and V_iV_j terms to remain in all further models

Focus on V_i, and V_iV_j terms *not* identified above: • candidates for elimination • assess confounding/precision for these variables

interaction terms in model
$$\Downarrow$$
final (confounding) step
difficult—subjective

Safest approach:
   keep all potential confounders in model:
      controls confounding but may lose
      precision

Confounding—general procedure:

$\widehat{OR}$ change?

gold standard   vs.   model without one
   model   or more V_i and V_iV_j

(1) Identify subsets so that
         $\widehat{OR}_{\text{gold standard}} \approx \widehat{OR}_{\text{subset}}$
(2) Control for largest gain in precision

difficult when there is interaction

In the second step of the flow diagram, we apply the Hierarchy Principle to identify those V_i and V_iV_j terms that are lower-order components of those interaction terms found significant. Such lower-order components must remain in all further models considered.

In the final step of the flow diagram, we focus on only those V_i and V_iV_j terms not identified by the Hierarchy Principle. These terms are **candidates** to be dropped from the model as nonconfounders. For those variables identified as candidates for elimination, we then **assess confounding followed by** consideration of **precision.**

If the model contains interaction terms, the final (confounding) step is difficult to carry out and requires subjectivity in deciding which variables can be eliminated as nonconfounders. We will illustrate such difficulties by the example below.

To avoid making subjective decisions, the safest approach is to keep all potential confounders in the model, whether or not they are eligible to be dropped. This will ensure the proper control of confounding but has the potential drawback of not giving as precise an odds ratio estimate as possible from some smaller subset of confounders.

In assessing confounding when there are interaction terms, the general procedure is analogous to when there is no interaction. We assess whether the estimated odds ratio changes from the gold standard model when compared to a model without one or more of the eligible V_i's and V_iV_j's.

More specifically, we carry out the following two steps:

(1) Identify those subsets of V_i's and V_iV_j's giving approximately the same odds ratio estimate as the gold standard.
(2) Control for that subset which gives the largest gain in precision.

$$\text{Interaction}: \widehat{OR} = \exp\!\left(\hat{\beta} + \sum \hat{\delta}_j W_j\right)$$

$$\hat{\beta} \text{ and } \hat{\delta}_j \text{ nonzero}$$

$$\text{no interaction}: \widehat{OR} = \exp\!\left(\hat{\beta}\right)$$

If the model contains interaction terms, the first step is difficult in practice. The odds ratio expression, as shown here, involves two or more coefficients, including one or more nonzero δ "hats." In contrast, when there is no interaction, the odds ratio involves the single coefficient β "hat."

Coefficients change when potential confounders dropped:

• meaningful change?
• subjective?

It is likely that at least one or more of the β "hat" and δ "hat" coefficients will change somewhat when potential confounders are dropped from the model. To evaluate how much of a change is a **meaningful** change when considering the collection of coefficients in the odds ratio formula is quite **subjective.** This will be illustrated by the example.

EXAMPLE

Variables in initial model:
$E, V_1, V_2, V_3, V_4 = V_1V_2$
$EV_1, EV_2, EV_3, EV_4 = EV_1V_2$

Suppose $EV_4\,(= EV_1V_2)$
significant

As an example, suppose our initial model contains E, four V's, namely, V_1, V_2, V_3, and $V_4 = V_1V_2$, and four EV's, namely, EV_1, EV_2, EV_3, and EV_4. Note that EV_4 alternatively can be considered as a three-factor product term as it is of the form EV_1V_2.

Suppose also that because EV_4 is a three-factor product term, it is tested first, after all the other variables are forced into the model. Further, suppose that this test is significant, so that the term EV_4 is to be retained in all further models considered.

Hierarchy Principle:

EV_1 and EV_2 retained in all further models
EV_3 candidate to be dropped

Because of the Hierarchy Principle, then, we must retain EV_1 and EV_2 in all further models as these two terms are components of EV_1V_2. This leaves EV_3 as the only remaining two-factor interaction candidate to be dropped if not significant.

Test for EV_3 (LR or Wald test)

To test for EV_3, we can do either a likelihood ratio test or a Wald test for the addition of EV_3 to a model after E, V_1, V_2, V_3, $V_4 = V_1V_2$, EV_1, EV_2 and EV_4 are forced into the model.

V_1, V_2, V_3, V_4 (**all** potential confounders) forced into model during interaction stage

Note that all four potential confounders—V_1 through V_4—are forced into the model here because we are at the interaction stage so far, and we have not yet addressed confounding in this example.

EXAMPLE (continued)

LR test for EV_3: Compare **full model** containing

$$E, \underbrace{V_1, V_2, V_3, V_4,}_{V\text{'s}} \underbrace{EV_1, EV_2, EV_3, EV_4}_{EV\text{'s}}$$

with **reduced model** containing

$$E, V_1, V_2, V_3, V_4, \underbrace{EV_1, EV_2, EV_4}_{\text{without } EV_3}$$

$$\text{LR} = \left(-2 \ln \hat{L}_{\text{reduced}}\right) - \left(-2 \ln \hat{L}_{\text{full}}\right)$$

is $\chi^2_{1\text{df}}$ under $\text{H}_0 : \beta_{EV_3} = 0$ **in full model**

Suppose EV_3 *not* significant
$\Downarrow$
model after interaction assessment:

$$E, \overbrace{(V_1, V_2, V_3, V_4)}, EV_1, EV_2, EV_4$$
where $V_4 = V_1 V_2$ $\qquad$ potential
$\qquad\qquad\qquad\qquad$ confounders

Hierarchy Principle:
$\quad$ identify V's not eligible to be
$\quad$ dropped—lower-order components

$EV_1 V_2$ significant
$\Downarrow$ Hierarchy Principle
Retain V_1, V_2, and $V_4 = V_1 V_2$
Only V_3 eligible to be dropped

$$\widehat{\text{OR}}_{V_1, V_2, V_3, V_4} \overset{?}{\neq} \widehat{\text{OR}}_{V_1, V_2, V_4}$$
$$\underset{\text{excludes } V_3}{\uparrow}$$

The likelihood ratio test for the significance of EV_3 compares a "full" model containing E, the four V's, EV_1, EV_2, EV_3, and EV_4 with a reduced model which eliminates EV_3 from the full model.

The LR statistic is given by the difference in the log likelihood statistics for the full and reduced models. This statistic has a chi-square distribution with one degree of freedom under the null hypothesis that the coefficient of the EV_3 term is 0 in our full model at this stage.

Suppose that when we carry out the LR test for this example, we find that the EV_3 term is not significant. Thus, at the end of the interaction assessment stage, we are left with a model that contains E, the four V's, EV_1, EV_2, and EV_4. We are now ready to assess confounding for this example.

Our initial model contained four potential confounders, namely, V_1 through V_4, where V_4 is the product term V_1 times V_2. Because of the Hierarchy Principle, some of these terms are not eligible to be dropped from the model, namely, the lower-order components of higher-order product terms remaining in the model.

In particular, because $EV_1 V_2$ has been found significant, we must retain in all further models the lower-order components V_1, V_2, and $V_1 V_2$, which equals V_4. This leaves V_3 as the only remaining potential confounder that is eligible to be dropped from the model as a possible nonconfounder.

To evaluate whether V_3 can be dropped from the model as a nonconfounder, we consider whether the odds ratio for the model which controls for all four potential confounders, including V_3, plus previously retained interaction terms, is meaningfully different from the odds ratio that controls for previously retained variables but excludes V_3.

EXAMPLE (continued)

$$\widehat{OR}_{V_1, V_2, V_3, V_4} = \exp\left(\hat{\beta} + \hat{\delta}_1 V_1 + \hat{\delta}_2 V_2 + \hat{\delta}_4 V_4\right),$$

where $\hat{\delta}_1$, $\hat{\delta}_2$, and $\hat{\delta}_4$ are coefficients of EV_1, EV_2, and $EV_4 = EV_1V_2$

$$\widehat{OR} = \exp\left(\hat{\beta} + \hat{\delta}_1 V_1 + \hat{\delta}_2 V_2 + \hat{\delta}_4 V_4\right)$$

$\widehat{OR}$ differs for different specifications of V_1, V_2, V_4

gold standard $\widehat{OR}$,
- controls for all potential confounders
- gives baseline $\widehat{OR}$

$$\widehat{OR}^* = \exp\left(\hat{\beta}^* + \hat{\delta}_1^* V_1 + \hat{\delta}_2^* V_2 + \hat{\delta}_4^* V_4\right),$$

where $\hat{\beta}^*$, $\hat{\delta}_1^*$, $\hat{\delta}_2^*$, $\hat{\delta}_4^*$ are coefficients in model without V_3

Model without V_3:
  $E, V_1, V_2, V_4, EV_1, EV_2, EV_4$

Model with V_3:
  $E, V_1, V_2 \, \widehat{V_3} \, V_4, EV_1, EV_2, EV_4$

Possible that
$\hat{\beta} \neq \hat{\beta}^*$, $\hat{\delta}_1 \neq \hat{\delta}_1^*$, $\hat{\delta}_2 \neq \hat{\delta}_2^*$, $\hat{\delta}_4 \neq \hat{\delta}_4^*$

The odds ratio that controls for all four potential confounders plus retained interaction terms is given by the expression shown here. This expression gives a formula for calculating numerical values for the odds ratio. This formula contains the coefficients β "hat," δ_1 "hat," δ_2 "hat," and δ_4 "hat," but also requires specification of three effect modifiers—namely, V_1, V_2, and V_4, which are in the model as product terms with E.

The numerical value computed for the odds ratio will differ depending on the values specified for the effect modifiers V_1, V_2, and V_4. This should not be surprising because the presence of interaction terms in the model means that the value of the odds ratio differs for different values of the effect modifiers.

The above odds ratio is the **gold standard** odds ratio expression for our example. This odds ratio controls for all potential confounders being considered, and it provides baseline odds ratio values to which all other odds ratio computations obtained from dropping candidate confounders can be compared.

The odds ratio that controls for previously retained variables but excludes the control of V_3 is given by the expression shown here. Note that this expression is essentially of the same form as the gold standard odds ratio. In particular, both expressions involve the coefficient of the exposure variable and the same set of effect modifiers.

However, the estimated coefficients for this odds ratio are denoted with an asterisk (*) to indicate that these estimates may differ from the corresponding estimates for the gold standard. This is because the model that excludes V_3 contains a different set of variables and, consequently, may result in different estimated coefficients for those variables in common to both models.

In other words, because the gold standard model contains V_3, whereas the model for the asterisked odds ratio does not contain V_3, it is possible that β "hat" will differ from β* "hat," and that the δ "hats" will differ from the δ* "hats."

EXAMPLE (continued)

Meaningful difference?

gold standard model:

$$\widehat{OR} = \exp\left(\hat{\beta} + \hat{\delta}_1 V_1 + \hat{\delta}_2 V_2 + \hat{\delta}_4 V_4\right)$$

model without V_3:

$$\widehat{OR}^* = \exp\left(\hat{\beta}^* + \hat{\delta}_1^* V_1 + \hat{\delta}_2^* V_2 + \hat{\delta}_4^* V_4\right)$$

$$\left(\hat{\beta}, \hat{\delta}_1, \hat{\delta}_2, \hat{\delta}_4,\right) \text{vs.} \left(\hat{\beta}^*, \hat{\delta}_1^*, \hat{\delta}_2^*, \hat{\delta}_4^*\right)$$

Difference?

Yes $\Rightarrow$ V_3 confounder; cannot eliminate V_3

No $\Rightarrow$ V_3 not confounder; drop V_3 if precision gain

Difficult approach:
• four coefficients to compare;
• coefficients likely to change

Overall decision required about change in $\hat{\beta}, \hat{\delta}_1, \hat{\delta}_2, \hat{\delta}_4$

More subjective than when no interaction (only $\hat{\beta}$)

To assess confounding here, we must determine whether there is a meaningful difference between the gold standard and asterisked odds ratio expressions. There are two alternative ways to do this.

One way is to compare corresponding estimated coefficients in the odds ratio expression, and then to make a decision whether there is a meaningful difference in one or more of these coefficients.

If we decide **yes,** that there is a difference, we then conclude that there is confounding due to V_3, so that we cannot eliminate V_3 from the model. If, on the other hand, we decide **no,** that corresponding coefficients are not different, we then conclude that we do not need to control for the confounding effects of V_3. In this case, we may consider dropping V_3 from the model if we can gain precision by doing so.

Unfortunately, this approach for assessing confounding is difficult in practice. In particular, in this example, the odds ratio expression involves four coefficients, and it is likely that at least one or more of these will change somewhat when one or more potential confounders are dropped from the model.

To evaluate whether there is a meaningful change in the odds ratio therefore requires an overall decision as to whether the collection of four coefficients, β "hat" and three δ "hats," in the odds ratio expression meaningfully change. This is a more subjective decision than for the no interaction situation when β "hat" is the only coefficient to be monitored.

EXAMPLE (continued)

$$\underbrace{\widehat{OR} = \exp\left(\hat{\beta} + \hat{\delta}_1 V_1 + \hat{\delta}_2 V_2 + \hat{\delta}_4 V_4\right)}_{\text{linear function}}$$

$\hat{\beta}, \hat{\delta}_1, \hat{\delta}_2, \hat{\delta}_4$ on log odds ratio scale; but odds ratio scale is clinically relevant

Log odds ratio scale:

$\hat{\beta} = -12.68$ vs. $\hat{\beta}^* = -12.72$

$\hat{\delta}_1 = 0.0691$ vs. $\hat{\delta}_1^* = 0.0696$

Odds ratio scale:

calculate $\widehat{OR} = \exp\left(\hat{\beta} + \sum \delta_j W_j\right)$

for different choices of W_j

Gold standard OR:

$\widehat{OR} = \exp\left(\hat{\beta} + \hat{\delta}_1 V_1 + \hat{\delta}_2 V_2 + \hat{\delta}_4 V_4\right)$
where $V_4 = V_1 V_2$

Specify V_1 and V_2 to get OR:

	$V_1=20$	$V_1=30$	$V_1=40$
$V_2=100$	$\widehat{OR}$	$\widehat{OR}$	$\widehat{OR}$
$V_2=200$	$\widehat{OR}$	$\widehat{OR}$	$\widehat{OR}$

Model without V_3:

$\widehat{OR}^* = \exp\left(\hat{\beta}^* + \hat{\delta}_1^* V_1 + \hat{\delta}_2^* V_2 + \hat{\delta}_4^* V_4\right)$

	$V_1=20$	$V_1=30$	$V_1=40$
$V_2=100$	$\widehat{OR}^*$	$\widehat{OR}^*$	$\widehat{OR}^*$
$V_2=200$	$\widehat{OR}^*$	$\widehat{OR}^*$	$\widehat{OR}^*$

Compare tables of

$\widehat{OR}$s vs. $\widehat{OR}^*$s
gold standard   model without V_3

Moreover, because the odds ratio expression involves the exponential of a linear function of the four coefficients, these coefficients are on a log odds ratio scale rather than an odds ratio scale. Using a log scale to judge the meaningfulness of a change is not as clinically relevant as using the odds ratio scale.

For example, a change in β "hat" from -12.68 to -12.72 and a change in δ_1 "hat" from 0.0691 to 0.0696 are not easy to interpret as clinically meaningful because these values are on a log odds ratio scale.

A more interpretable approach, therefore, is to view such changes on the odds ratio scale. This involves calculating numerical values for the odds ratio by substituting into the odds ratio expression different choices of the values for the effect modifiers W_j.

Thus, to calculate an odds ratio value from the gold standard formula shown here, which controls for all four potential confounders, we would need to specify values for the effect modifiers V_1, V_2, and V_4, where V_4 equals $V_1 V_2$. For different choices of V_1 and V_2, we would then obtain different odds ratio values. This information can be summarized in a table or graph of odds ratios which consider the different specifications of the effect modifiers. A sample table is shown here.

To assess confounding on an odds ratio scale, we would then compute a similar table or graph which would consider odds ratio values for a model which drops one or more eligible V variables. In our example, because the only eligible variable is V_3, we, therefore, need to obtain an odds ratio table or graph for the model that does not contain V_3. A sample table of OR^* values is shown here.

Thus, to assess whether we need to control for confounding from V_3, we need to compare two tables of odds ratios, one for the gold standard and the other for the model which does not contain V_3.

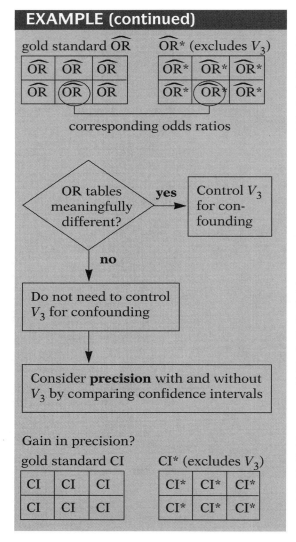

If, looking at these two tables collectively, we find that **yes**, there is one or more meaningful difference in corresponding odds ratios, we would conclude that the variable V_3 needs to be controlled for confounding. In contrast, if we decide that **no**, the two tables are not meaningfully different, we can conclude that variable V_3 does not need to be controlled for confounding.

If the decision is made that V_3 does not need to be controlled for confounding reasons, we still may wish to control for V_3 because of precision reasons. That is, we can compare confidence intervals for corresponding odds ratios from each table to determine whether we gain or lose precision depending on whether or not V_3 is in the model.

In other words, to assess whether there is a gain in precision from dropping V_3 from the model, we need to make an overall comparison of two tables of confidence intervals for odds ratio estimates obtained when V_3 is in and out of the model.

EXAMPLE (continued)

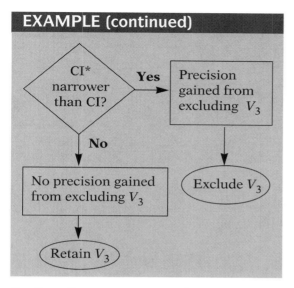

If, overall, we decide that **yes**, the asterisked confidence intervals, which exclude V_3, are narrower than those for the gold standard table, we would conclude that precision is gained from excluding V_3 from the model. Otherwise, if we decide **no,** then we conclude that no meaningful precision is gained from dropping V_3, and so we retain this variable in our final model.

Confounding assessment when interaction present (summary):

- compare tables of ORs and CIs
- subjective—debatable
- safest decision—control for all potential counfounders

Thus, we see that when there is interaction and we want to assess both confounding and precision, we must compare tables of odds ratio point estimates followed by tables of odds ratio confidence intervals. Such comparisons are quite subjective and, therefore, debatable in practice. That is why the safest decision is to control for all potential confounders even if some V's are candidates to be dropped.

V. The Evans County Example Continued

EXAMPLE

Evans County Heart Disease Study

$n = 609$ white males
9-year follow-up

$D = \text{CHD}_{(0,\,1)}$
$E = \text{CAT}_{(0,\,1)}$

$C\text{'s}: \underset{\text{continuous}}{\underline{\text{AGE, CHL}}} \ \underset{(0,\,1)}{\underline{\text{SMK, ECG, HPT}}}$

We now review the interaction and confounding assessment recommendations by returning to the Evans County Heart Disease Study Data that we have considered in the previous chapters.

Recall that the study data involves 609 white males followed for 9 years to determine CHD status. The exposure variable is catecholamine level (CAT), and the C variables considered for control are AGE, cholesterol (CHL), smoking status (SMK), electrocardiogram abnormality status (ECG), and hypertension status (HPT). The variables AGE and CHL are treated continuously, whereas SMK, ECG, and HPT are $(0, 1)$ variables.

EXAMPLE (continued)

Initial E, V, W model:

$$\text{logit } P(\mathbf{X}) = \alpha + \beta CAT + \sum_{i=1}^{5} \gamma_i V_i + E \sum_{j=1}^{5} \delta_j W_j$$

where V's = C's = W's
HWF model because
EV_i in model
$\Downarrow$
E and V_i in model

Highest order in model: EV_i
    no EV_iV_j or V_iV_j terms

Next step:
    interaction assessment using
    backward elimination

Backward elimination:

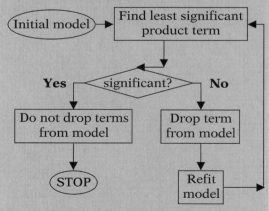

Interaction results:

eliminated	remaining
CAT × AGE	CAT × CHL
CAT × SMK	CAT × HPT
CAT × ECG	

In the variable specification stage of our strategy, we choose an initial E, V, W model, shown here, containing the exposure variable CAT, five V's which are the C's themselves, and five W's which are also the C's themselves and which go into the model as product terms with the exposure CAT.

This initial model is HWF because the lower-order components of any EV_i term, namely, E and V_i, are contained in the model.

Note also that the highest-order terms in this model are two-factor product terms of the form EV_i. Thus, we are not considering more complicated three-factor product terms of the form EV_iV_j nor V_i terms which are of the form V_iV_j.

The next step in our modeling strategy is to consider eliminating unnecessary interaction terms. To do this, we use a backward elimination procedure to remove variables. For interaction terms, we proceed by eliminating product terms one at a time.

The flow for our backward procedure begins with the initial model and then identifies the least significant product term. We then ask, "Is this term significant?" If our answer is **no,** we eliminate this term from the model. The model is then refitted using the remaining terms. The least significant of these remaining terms is then considered for elimination.

This process continues until our answer to the significance question in the flow diagram is **yes.** If so, the least significant term is significant in some refitted model. Then, no further terms can be eliminated, and our process must stop.

For our initial Evans County model, the backward procedure allows us to eliminate the product terms of CAT × AGE, CAT × SMK, and CAT × ECG. The remaining interaction terms are CAT × CHL and CAT × HPT.

EXAMPLE (continued)

Printout:

Variable	Coefficient	S.E.	Chi sq	P
Intercept	−4.0474	1.2549	10.40	0.0013
CAT	−12.6809	3.1042	16.69	0.0000
AGE	0.0349	0.0161	4.69	0.0303
CHL	−0.0055	0.0042	1.70	0.1917
ECG	0.3665	0.3278	1.25	0.2635
SMK	0.7735	0.3272	5.59	0.0181
HPT	1.0468	0.3316	9.96	0.0016
CH	−2.3299	0.7422	(9.85)	(0.0017)
CC	0.0691	0.0143	23.18	0.0000

V's { AGE, CHL, ECG, SMK, HPT }

— W's —

CH = CAT × HPT and CC = CAT × CHL remain in all further models

Confounding assessment:
Step 1. Variables in model:

CAT, $\underbrace{\text{AGE, CHL, SMK, ECG, HPT}}_{V\text{'s}}$

$\underbrace{\text{CAT×CHL, CAT×HPT,}}_{EV\text{'s}}$

All five V's still in model after interaction

Hierarchy Principle:

- determine V's that cannot be eliminated
- all lower-order components of significant product terms remain

CAT × CHL significant ⇒ CAT and CHL components

CAT × HPT significant ⇒ CAT and HPT components

A summary of the printout for the model remaining after interaction assessment is shown here. In this model, the two interaction terms are CH equals CAT × HPT and CC equals CAT × CHL. The least significant of these two terms is CH because the Wald statistic for this term is given by the chi-square value of 9.85, which is less significant than the chi-square value of 23.18 for the CC term.

The P-value for the CH term is 0.0017, so that this term is significant at well below the 1% level. Consequently, we cannot drop CH from the model, so that all further models must contain the two product terms CH and CC.

We are now ready to consider the confounding assessment stage of the modeling strategy. The first step in this stage is to identify all variables remaining in the model after the interaction stage. These are CAT, all five V variables, and the two product terms CAT × CHL and CAT × HPT.

The reason why the model contains all five V's at this point is that we have only completed interaction assessment and have not yet begun to address confounding to evaluate which of the V's can be eliminated from the model.

The next step is to apply the Hierarchy Principle to determine which V variables cannot be eliminated from further models considered.

The Hierarchy Principle requires all lower-order components of significant product terms to remain in all further models.

The two significant product terms in our model are CAT × CHL and CAT × HPT. The lower-order components of CAT × CHL are CAT and CHL. The lower-order components of CAT × HPT are CAT and HPT.

EXAMPLE (continued)

Thus, retain CAT, CHL, and HPT in all further models

Candidates for elimination:
  AGE, SMK, ECG

Assessing confounding:
  do coefficients in $\widehat{OR}$ expression change?

$\widehat{OR} = \exp\left(\hat{\beta} + \hat{\delta}_1 CHL + \hat{\delta}_2 HPT\right),$
where

$\hat{\beta} = $ coefficient of CAT

$\hat{\delta}_1 = $ coefficient of CC $=$ CAT $\times$ CHL

$\hat{\delta}_2 = $ coefficient of CH $=$ CAT $\times$ HPT

Gold standard $\widehat{OR}$ (all V's):

$\widehat{OR} = \exp\left(\hat{\beta} + \hat{\delta}_1 CHL + \hat{\delta}_2 HPT\right),$
where

$\hat{\beta} = -12.6809,\ \hat{\delta}_1 = 0.0691,\ \hat{\delta}_2 = -2.3299$

V_i in model	$\hat{\beta}$	$\hat{\delta}_1$	$\hat{\delta}_2$
All five V variables	-12.6809	0.0691	-2.3299
CHL, HPT, AGE, ECG	-12.7187	0.0696	-2.3809
CHL, HPT, AGE, SMK	-12.8367	0.0706	-2.3316
CHL, HPT, ECG, SMK	-12.5611	0.0696	-2.2059
CHL, HPT, AGE	-12.7787	0.0760	-2.3769
CHL, HPT, ECG	-12.5767	0.0703	-2.2561
CHL, HPT, SMK	-12.7131	0.0711	-2.2188
CHL, HPT	-12.7324	0.0712	-2.2584

Because CAT is the exposure variable, we must leave CAT in all further models regardless of the Hierarchy Principle. In addition, CHL and HPT are the two V's that must remain in all further models.

This leaves the V variables AGE, SMK, and ECG as still being candidates for elimination as possible nonconfounders.

As described earlier, one approach to assessing whether AGE, SMK, and ECG are nonconfounders is to determine whether the coefficients in the odds ratio expression for the CAT, CHD relationship change meaningfully as we drop one or more of the candidate terms AGE, SMK, and ECG.

The odds ratio expression for the CAT, CHD relationship is shown here. This expression contains β "hat," the coefficient of the CAT variable, plus two terms of the form δ "hat" times W, where the W's are the effect modifiers CHL and HPT that remain as a result of interaction assessment.

The gold standard odds ratio expression is derived from the model remaining after interaction assessment. This model controls for all potential confounders, that is, the V's, in the initial model. For the Evans County data, the coefficients in this odds ratio, which are obtained from the printout above, are β "hat" equals -12.6809, δ_1 "hat" equals 0.0691, and δ_2 "hat" equals -2.3299.

The table shown here provides the odds ratio coefficients β "hat," δ_1 "hat," and δ_2 "hat" for different subsets of AGE, SMK, and ECG in the model. The first row of coefficients is for the gold standard model, which contains all five V's. The next row shows the coefficients obtained when SMK is dropped from the model, and so on down to the last row which shows the coefficients obtained when AGE, SMK, and ECG are simultaneously removed from the model so that only CHL and HPT are controlled.

EXAMPLE (continued)

Coefficients change somewhat. No radical change

In scanning the above table, it is seen for each coefficient separately (that is, by looking at the values in a given column) that the estimated values change somewhat as different subsets of AGE, SMK, and ECG are dropped. However, there does not appear to be a radical change in any coefficient.

Meaningful differences in $\widehat{OR}$?

- coefficients on log odds ratio scale
- more appropriate: odds ratio scale

Nevertheless, it is not clear whether there is sufficient change in any coefficient to indicate meaningful differences in odds ratio values. Assessing the effect of a change in coefficients on odds ratio values is difficult because the coefficients are on the log odds ratio scale. It is more appropriate to make our assessment of confounding using odds ratio values rather than log odds ratio values.

$$\widehat{OR} = \exp\left(\hat{\beta} + \hat{\delta}_1 CHL + \hat{\delta}_2 HPT\right)$$

Specify values of effect modifiers
Obtain summary table of ORs

To obtain numerical values for the odds ratio for a given model, we must specify values of the effect modifiers in the odds ratio expression. Different specifications will lead to different odds ratios. Thus, for a given model, we must consider a summary table or graph that describes the different odds ratio values that are calculated.

Compare
gold standard vs. other models
using   (without V's)
odds ratio tables or graphs

To compare the odds ratios for two different models, say the gold standard model with the model that deletes one or more eligible V variables, we must compare corresponding odds ratio tables or graphs.

Evans County example:
gold standard
vs.
model without AGE, SMK, and ECG

As an illustration using the Evans County data, we compare odds ratio values computed from the gold standard model with values computed from the model which deletes the three eligible variables AGE, SMK, and ECG.

Gold standard $\widehat{OR}$:
$$\widehat{OR} = \exp\left(-12.6809 + 0.0691 CHL - 2.3299 HPT\right)$$

	HPT = 0	HPT = 1
CHL = 200	$\widehat{OR}$ = 3.12	$\widehat{OR}$ = 0.30
CHL = 220	$\widehat{OR}$ = 12.44	$\widehat{OR}$ = 1.21
CHL = 240	$\widehat{OR}$ = 49.56	$\widehat{OR}$ = 4.82

CHL = 200, HPT = 0 $\Rightarrow \widehat{OR} = 3.12$
CHL = 220, HPT = 1 $\Rightarrow \widehat{OR} = 1.21$

The table shown here gives odds ratio values for the gold standard model, which contains all five V variables, the exposure variable CAT, and the two interaction terms CAT × CHL and CAT × HPT. In this table, we have specified three different row values for CHL, namely, 200, 220, and 240, and two column values for HPT, namely, 0 and 1. For each combination of CHL and HPT values, we thus get a different odds ratio.

EXAMPLE (continued)

CHL = 200, HPT = 0 $\Rightarrow \widehat{OR}$ = $\boxed{3.12}$
CHL = 220, HPT = 1 $\Rightarrow \widehat{OR}$ = $\boxed{1.21}$

$\widehat{OR}$ with AGE, SMK, ECG deleted:

$\widehat{OR}^* = \exp(-12.7324 + 0.0712CHL - 2.2584HPT)$

	HPT = 0	HPT = 1
CHL = 200	$\widehat{OR}^*$ = 4.56	$\widehat{OR}^*$ = 0.48
CHL = 220	$\widehat{OR}^*$ = 18.96	$\widehat{OR}^*$ = 1.91
CHL = 240	$\widehat{OR}^*$ = 78.82	$\widehat{OR}^*$ = 8.24

Gold standard $\widehat{OR}$: $\widehat{OR}^*$ w/o AGE, SMK, ECG

	HPT=0	HPT=1	HPT=0	HPT=1
CHL=200	$\boxed{3.12}$	0.30	$\boxed{4.56}$	0.48
CHL=220	12.44	$\boxed{1.21}$	18.96	$\boxed{1.98}$
CHL=240	49.56	4.82	78.82	8.24

Cannot simultaneously drop AGE, SMK, and ECG from model

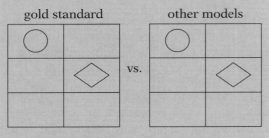

gold standard   other models

vs.

Other models: delete AGE and SMK or delete AGE and ECG, etc.

Result: cannot drop AGE, SMK, or ECG

Final model:

E: CAT
five V's: CHL, HPT, AGE, SMK, ECG
two interactions: CAT × CHL, CAT × HPT

For example, if CHL equals 200 and HPT equals 0, the computed odds ratio is 3.12, whereas if CHL equals 220 and HPT equals 1, the computed odds ratio is 1.21.

The table shown here gives odds ratio values, indicated by "asterisked" OR "hats," for a model that deletes the three eligible V variables, AGE, SMK, and ECG. As with the gold standard model, the odds ratio expression involves the same two effect modifiers CHL and HPT, and the table shown here considers the same combination of CHL and HPT values.

If we compare corresponding odds ratios in the two tables, we can see sufficient discrepancies.

For example, when CHL equals 200 and HPT equals 0, the odds ratio is 3.12 in the gold standard model, but is 4.56 when AGE, SMK, and ECG are deleted. Also, when CHL equals 220 and HPT equals 1, the corresponding odds ratios are 1.21 and 1.98.

Thus, because the two tables of odds ratios differ appreciably, we cannot simultaneously drop AGE, SMK, and ECG from the model.

Similar comparisons can be made by comparing the gold standard odds ratio with odds ratios obtained by deleting other subsets, for example, AGE and SMK together, or AGE and ECG together, and so on. All such comparisons show sufficient discrepancies in corresponding odds ratios. Thus, we cannot drop any of the three eligible variables from the model.

We conclude that all five V variables need to be controlled, so that the final model contains the exposure variable CAT, the five V variables, and the interaction variables involving CHL and HPT.

EXAMPLE (continued)

No need to consider precision in this example:
   compare tables of CIs—subjective

Confounding and precision difficult if interaction (subjective)

Caution: do not sacrifice validity for minor gain in precision

Summary result for final model:

	Table of $\widehat{OR}$			Table of 95% CIs	
CHL	HPT=0	HPT=1		HPT=0	HPT=1
200	3.12	0.30		(0.90, 10.96)	(0.10, 0.90)
220	12.44	1.21		(3.67, 42.57)	(0.48, 3.08)
240	49.56	4.82		(11.79, 210.05)	(1.63, 14.39)

Use to draw meaningful conclusions

$$\text{CHL} \nearrow \Rightarrow \widehat{OR}_{\text{CAT, CHD}} \nearrow$$

$$\text{CHL fixed}: \widehat{OR}_{\substack{\text{CAT, CHD} \\ \text{HPT=0}}} > \widehat{OR}_{\substack{\text{CAT, CHD} \\ \text{HPT=1}}}$$

All CIs are wide

Note that because we cannot drop either of the variables AGE, SMK, or ECG as nonconfounders, we do not need to consider possible gain in precision from deleting nonconfounders. If precision were considered, we would compare tables of confidence intervals for different models. As with confounding assessment, such comparisons are largely subjective.

This example illustrates why we will find it difficult to assess confounding and precision if our model contains interaction terms. In such a case, any decision to delete possible nonconfounders is largely subjective. Therefore, we urge caution when deleting variables from our model in order to avoid sacrificing validity in exchange for what is typically only a minor gain in precision.

To conclude this example, we point out that, using the final model, a summary of the results of the analysis can be made in terms of the table of odds ratios and the corresponding table of confidence intervals.

Both tables are shown here. The investigator must use this information to draw meaningful conclusions about the relationship under study. In particular, the nature of the interaction can be described in terms of the point estimates and confidence intervals.

For example, as CHL increases, the odds ratio for the effect of CAT on CHD increases. Also, for fixed CHL, this odds ratio is higher when HPT is 0 than when HPT equals 1. Unfortunately, all confidence intervals are quite wide, indicating that the point estimates obtained are quite unstable.

EXAMPLE (continued)

Tests of significance:

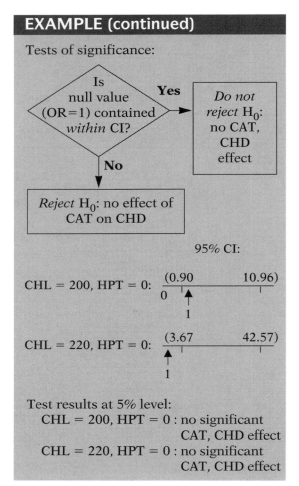

Furthermore, tests of significance can be carried out using the confidence intervals. To do this, one must determine whether or not the null value of the odds ratio, namely, 1, is contained within the confidence limits. If so, we do not reject, for a given CHL, HPT combination, the null hypothesis of no effect of CAT on CHD. If the value 1 lies outside the confidence limits, we would reject the null hypothesis of no effect.

For example, when CHL equals 200 and HPT equals 0, the value of 1 is contained within the limits 0.90 and 10.96 of the 95% confidence interval. However, when CHL equals 220 and HPT equals 0, the value of 1 is not contained within the limits 3.67 and 42.57.

Test results at 5% level:
  CHL = 200, HPT = 0 : no significant
              CAT, CHD effect
  CHL = 220, HPT = 0 : no significant
              CAT, CHD effect

Thus, when CHL equals 200 and HPT equals 0, there is no significant CAT, CHD effect, whereas when CHL equals 220 and HPT equals 0, the CAT, CHD effect is significant at the 5% level.

Tests based on CIs are *two-tailed*

In EPID, most tests of *E–D* relationship are *one-tailed*

Note that tests based on confidence intervals are two-tailed tests. One-tailed tests are more common in epidemiology for testing the effect of an exposure on disease.

One-tailed tests:
  use large sample

$$Z = \frac{\text{estimate}}{\text{standard error}}$$

When there is interaction, one-tailed tests can be obtained by using the point estimates and their standard errors that go into the computation of the confidence interval. The point estimate divided by its standard error gives a large sample Z statistic which can be used to carry out a one-tailed test.

SUMMARY

Chapter 6

- overall guidelines for three stages
- focus: variable specification
- HWF model

Chapter 7

Focus: interaction and confounding
          assessment

Interaction: use hierarchical backward
                elimination

Use **Hierarchy Principle** to identify lower-order components that cannot be deleted (EV's, V_i's, and V_iV_j's)

Confounding: *no* statistical testing:
    Compare whether $\widehat{OR}$ meaningfully changes when V's are deleted

Drop nonconfounders if precision is gained by examining CIs

No interaction: assess confounding by monitoring changes in $\hat{\beta}$, the coefficient of E

A brief summary of this presentation is now given. This has been the second of two chapters on modeling strategy. In Chapter 6, we gave overall guidelines for three stages, namely, variable specification, interaction assessment, and confounding assessment, with consideration of precision. Our primary focus was the variable specification stage, and an important requirement was that the initial model be hierarchically well formulated (HWF).

In this chapter, we have focused on the interaction and confounding assessment stages of our modeling strategy. We have described how interaction assessment follows a hierarchical backward elimination procedure, starting with assessing higher-order interaction terms followed by assessing lower-order interaction terms using statistical testing methods.

If certain interaction terms are significant, we use the Hierarchy Principle to identify all lower-order components of such terms, which cannot be deleted from any further model considered. This applies to lower-order interaction terms (that is, terms of the form EV) and to lower-order terms involving potential confounders of the form V_i or V_iV_j.

Confounding is assessed without the use of statistical testing. The procedure involves determining whether the estimated odds ratio meaningfully changes when eligible V variables are deleted from the model.

If some variables can be identified as nonconfounders, they may be dropped from the model provided their deletion leads to a gain in precision from examining confidence intervals.

If there is no interaction, the assessment of confounding is carried out by monitoring changes in the estimated coefficient of the exposure variable.

SUMMARY (*continued*)

Interaction present: compare tables of odds ratios and confidence intervals (subjective)

Interaction: Safe (for validity) to keep all V's in model

However, if there is interaction, the assessment of confounding is much more subjective because it typically requires the comparison of tables of odds ratio values. Similarly, assessing precision requires comparison of tables of confidence intervals.

Consequently, if there is interaction, it is typically safe for ensuring validity to keep all potential confounders in the model, even those that are candidates to be deleted as possible nonconfounders.

Chapters
1. Introduction
2. Special Cases

   •

   •

✓ 7. Interaction and Confounding Assessment
8. Analysis of Matched Data

This presentation is now complete. The reader may wish to review the detailed summary and to try the practice exercises and test that follow.

The next chapter concerns the use of logistic modeling to assess matched data.

Detailed Outline

Practice Exercises

A prevalence study of predictors of surgical wound infection in 265 hospitals throughout Australia collected data on 12,742 surgical patients (McLaws et al., *Med. J. of Australia*, 1988). For each patient, the following independent variables were determined: type of hospital (public or private), size of hospital (large or small), degree of contamination of surgical site (clean or contaminated), and age and sex of the patient. A logistic model was fit to this data to predict whether or not the patient developed a surgical wound infection during hospitalization. The abbreviated variable names and the manner in which the variables were coded in the model are described as follows:

Variable	Abbreviation	Coding
Type of hospital	HT	1 = public, 0 = private
Size of hospital	HS	1 = large, 0 = small
Degree of contamination	CT	1 = contaminated, 0 = clean
Age	AGE	Continuous
Sex	SEX	1 = female, 0 = male

1. Suppose the following initial model is specified for assessing the effect of type of hospital (HT), considered as the exposure variable, on the prevalence of surgical wound infection, controlling for the other four variables on the above list:

   logit $P(\mathbf{X}) = \alpha + \beta HT + \gamma_1 HS + \gamma_2 CT + \gamma_3 AGE + \gamma_4 SEX + \delta_1 HT \times AGE + \delta_2 HT \times SEX$.

   Describe how to test for the overall significance (a "chunk" test) of the interaction terms. In answering this, describe the null hypothesis, the full and reduced models, the form of the test statistic, and its distribution under the null hypothesis.

2. Using the model given in Exercise 1, describe briefly how to carry out a backward elimination procedure to assess interaction.

3. Briefly describe how to carry out interaction assessment for the model described in Exercise 1. (In answering this, it is suggested you make use of the tests described in Exercises 1 and 2.)

4. Suppose the interaction assessment stage for the model in Example 1 finds no significant interaction terms. What is the formula for the odds ratio for the effect of HT on the prevalence of surgical wound infection at the end of the interaction assessment stage? What V terms remain in the model at the end of interaction assessment? Describe how you would evaluate which of these V terms should be controlled as confounders.

5. Considering the scenario described in Exercise 4 (i.e., no interaction terms found significant), suppose you determine that the variables CT and AGE do not need to be controlled for confounding. Describe how you would consider whether dropping both variables will improve precision.

6. Suppose the interaction assessment stage finds that the interaction terms HT $\times$ AGE and HT $\times$ SEX are both significant. Based on this result, what is the formula for the odds ratio that describes the effect of HT on the prevalence of surgical wound infection?

7. For the scenario described in Example 6, and making use of the Hierarchy Principle, what V terms are eligible to be dropped as possible nonconfounders?

8. Describe briefly how you would assess confounding for the model considered in Exercises 6 and 7.

9. Suppose that the variable SEX is determined to be a *non*confounder, whereas all other V variables in the model (of Exercise 1) need to be controlled. Describe briefly how you would assess whether the variable SEX needs to be controlled for precision reasons.

10. What problems are associated with the assessment of confounding and precision described in Exercises 8 and 9?

Test

The following questions consider the use of logistic regression on data obtained from a matched case-control study of cervical cancer in 313 women from Sydney, Australia (Brock et al., *J. Nat. Cancer Inst.*, 1988). The outcome variable is cervical cancer status (1 = present, 0 = absent). The matching variables are age and socioeconomic status. Additional independent variables not matched on are smoking status, number of lifetime sexual partners, and age at first sexual intercourse. The independent variables are listed below together with their computer abbreviation and coding scheme.

Variable	Abbreviation	Coding
Smoking status	SMK	1 = ever, 0 = never
Number of sexual partners	NS	1 = 4+, 0 = 0–3
Age at first intercourse	AS	1= 20+, 0 = ≤19
Age of subject	AGE	Category matched
Socioeconomic status	SES	Category matched

Assume that at the end of the variable specification stage, the following E, V, W model has been defined as the initial model to be considered:

$$\text{logit } P(\mathbf{X}) = \alpha + \beta \text{SMK} + \sum \gamma_i^* V_i^* + \gamma_1 \text{NS} + \gamma_2 \text{AS} + \gamma_3 \text{NS} \times \text{AS}$$
$$+ \delta_1 \text{SMK} \times \text{NS} + \delta_2 \text{SMK} \times \text{AS} + \delta_3 \text{SMK} \times \text{NS} \times \text{AS},$$

where the V_i^* are dummy variables indicating matching strata, the γ_i^* are the coefficients of the V_i^* variables, SMK is the only exposure variable of interest, and the variables NS, AS, AGE, and SES are being considered for control.

1.　For the above model, which variables are interaction terms?
2.　For the above model, list the steps you would take to assess interaction using a hierarchically backward elimination approach.
3.　Assume that at the end of interaction assessment, the only interaction term found significant is the product term SMK × NS. What variables are left in the model at the end of the interaction stage? Which of the V variables in the model cannot be deleted from any further models considered? Explain briefly your answer to the latter question.
4.　Based on the scenario described in Question 3 (i.e., the only significant interaction term is SMK × NS), what is the expression for the odds ratio that describes the effect of SMK on cervical cancer status at the end of the interaction assessment stage?
5.　Based again on the scenario described in Question 3, what is the expression for the odds ratio that describes the effect of SMK on cervical cancer status if the variable NS × AS is dropped from the model that remains at the end of the interaction assessment stage?
6.　Based again on the scenario described in Question 3, how would you assess whether the variable NS × AS should be retained in the model? (In answering this question, consider both confounding and precision issues.)

7. Suppose the variable NS × AS is dropped from the model based on the scenario described in Question 13. Describe how you would assess confounding and precision for any other V terms still eligible to be deleted from the model after interaction assessment.

8. Suppose the final model obtained from the cervical cancer study data is given by the following printout results:

Variable	β	S.E.	Chi sq	P
SMK	1.9381	0.4312	20.20	0.0000
NS	1.4963	0.4372	11.71	0.0006
AS	−0.6811	0.3473	3.85	0.0499
SMK × NS	−1.1128	0.5997	3.44	0.0635

Describe briefly how you would use the above information to summarize the results of your study. (In your answer, you need only describe the information to be used rather than actually calculate numerical results.)

Answers to Practice Exercises

1. A "chunk" test for overall significance of interaction terms can be carried out using a likelihood ratio test that compares the initial (full) model with a reduced model under the null hypothesis of no interaction terms. The likelihood ratio test will be a chi-square test with two degrees of freedom (because two interaction terms are being tested simultaneously).

2. Using a backward elimination procedure, one first determines which of the two product terms HT × AGE and HT × SEX is the least significant in a model containing these terms and all main effect terms. If this least significant term is significant, then both interaction terms are retained in the model. If the least significant term is nonsignificant, it is then dropped from the model. The model is then refitted with the remaining product term and all main effects. In the refitted model, the remaining interaction term is tested for significance. If significant, it is retained; if not significant, it is dropped.

3. Interaction assessment would be carried out first using a "chunk" test for overall interaction as described in Exercise 1. If this test is not significant, one could drop both interaction terms from the model as being not significant overall. If the chunk test is significant, then backward elimination, as described in Exercise 2, can be carried out to decide if both interaction terms need to be retained or whether one of the terms can be dropped. Also, even if the chunk test is not significant, backward elimination may be carried out to determine whether a significant interaction term can still be found despite the chunk test results.

4. The odds ratio formula is given by exp(β), where β is the coefficient of the HT variable. All V variables remain in the model at the end of the interaction assessment stage. These are HS, CT, AGE, and SEX. To

evaluate which of these terms are confounders, one has to consider whether the odds ratio given by exp(β) changes as one or more of the V variables are dropped from the model. If, for example, HS and CT are dropped and exp(β) does not change from the (gold standard) model containing all V's, then HS and CT do not need to be controlled as confounders. Ideally, one should consider as candidates for control any subset of the four V variables that will give the same odds ratio as the gold standard.

5. If CT and AGE do not need to be controlled for confounding, then, to assess precision, we must look at the confidence intervals around the odds ratio for a model which contains neither CT nor AGE. If this confidence interval is meaningfully narrower than the corresponding confidence interval around the gold standard odds ratio, then precision is gained by dropping CT and AGE. Otherwise, even though these variables need not be controlled for confounding, they should be retained in the model if precision is not gained by dropping them.

6. The odds ratio formula is given by exp($\beta + \delta_1$AGE $+ \delta_2$SEX).

7. Using the Hierarchy Principle, CT and HS are eligible to be dropped as nonconfounders.

8. Drop CT, HS, or both CT and HS from the model and determine whether the coefficients β, δ_1, and δ_2 in the odds ratio expression change. Alternatively, determine whether the odds ratio itself changes by comparing tables of odds ratios for specified values of the effect modifiers AGE and SEX. If there is no change in coefficients and/or in odds ratio tables, then the variables dropped do not need to be controlled for confounding.

9. Drop SEX from the model and determine if the confidence interval around the odds ratio is wider than the corresponding confidence interval for the model which contains SEX. Because the odds ratio is defined by the expression exp($\beta + \delta_1$AGE $+ \delta_2$SEX), a table of confidence intervals for both the model without SEX and with SEX will need to be obtained by specifying different values for the effect modifiers AGE and SEX. To assess whether SEX needs to be controlled for precision reasons, one must compare these tables of confidence intervals. If the confidence intervals when SEX is not in the model are narrower in some overall sense than when SEX is in the model, precision is gained by dropping SEX. Otherwise, SEX should be controlled as precision is not gained when the SEX variable is removed.

10. Assessing confounding and precision in Exercises 8 and 9 requires subjective comparisons of either several regression coefficients, several odds ratios, or several confidence intervals. Such subjective comparisons are likely to lead to highly debatable conclusions, so that a safe course of action is to control for all V variables regardless of whether they are confounders or not.

8

Analysis of Matched Data Using Logistic Regression

Introduction

Our discussion of matching begins with a general description of the matching procedure and the basic features of matching. We then discuss how to use stratification to carry out a matched analysis. Our primary focus is on case-control studies. We then introduce the logistic model for matched data and describe the corresponding odds ratio formula. Finally, we illustrate the analysis of matched data using logistic regression with an application that involves matching as well as control variables not involved in matching.

Abbreviated Outline

The outline below gives the user a preview of this chapter. A detailed outline for review purposes follows the presentation.

Objectives   Upon completion of this chapter, the learner should be able to:

1. State or recognize the procedure used when carrying out matching in a given study.
2. State or recognize at least one advantage and one disadvantage of matching.
3. State or recognize when to match or not to match in a given study situation.
4. State or recognize why attaining validity is not a justification for matching.
5. State or recognize two equivalent ways to analyze matched data using stratification.
6. State or recognize the McNemar approach for analyzing pair-matched data.
7. State or recognize the general form of the logistic model for analyzing matched data as an E, V, W-type model.
8. State or recognize an appropriate logistic model for the analysis of a specified study situation involving matched data.
9. State how dummy or indicator variables are defined and used in the logistic model for matched data.
10. Outline a recommended strategy for the analysis of matched data using logistic regression.
11. Apply the recommended strategy as part of the analysis of matched data using logistic regression.

Presentation

I. Overview

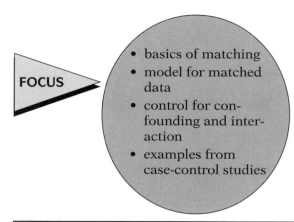

FOCUS

- basics of matching
- model for matched data
- control for confounding and interaction
- examples from case-control studies

This presentation describes how logistic regression may be used to analyze matched data. We describe the basic features of matching, and then focus on a general form of the logistic model for matched data that controls for confounding and interaction. We also provide examples of this model involving matched case-control data.

II. Basic Features of Matching

Study design procedure:
- select referent group
- comparable to index group on one or more "matching factors"

Matching is a procedure carried out at the design stage of a study which compares two or more groups. To match, we select a referent group for our study that is to be compared with the group of primary interest, called the index group. Matching is accomplished by constraining the referent group to be comparable to the index group on one or more risk factors, called "matching factors."

EXAMPLE

Matching factor = AGE

referent group constrained to have **same age structure** as index group

For example, if the matching factor is age, then matching on age would constrain the referent group to have essentially the same age structure as the index group.

Case-control study:
↑
our focus   referent = controls
             index = cases

Follow-up study:
referent = unexposed
index = exposed

In a case-control study, the referent group consists of the controls, which is compared to an index group of cases.

In a follow-up study, the referent group consists of unexposed subjects, which is compared to the index group of exposed subjects.

Henceforth in this presentation, we focus on case-control studies, but the model and methods described apply to follow-up studies also.

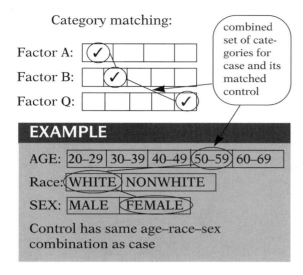

Category matching:

Factor A:

Factor B:

Factor Q:

combined set of categories for case and its matched control

EXAMPLE

AGE: 20–29 | 30–39 | 40–49 | 50–59 | 60–69

Race: WHITE | NONWHITE

SEX: MALE | FEMALE

Control has same age–race–sex combination as case

Case	No. of controls	Type
1	1	1–1 or pair matching
1	R	R-to1 (e.g., $R = 4 \rightarrow$ 4-to-1)

R may vary from case to case

e.g. $\begin{cases} R = 3 \text{ for some cases} \\ R = 2 \text{ for other cases} \\ R = 1 \text{ for other cases} \end{cases}$

Not always possible to find exactly R controls for each case

To match or **not to match**

Advantage
  Matching can be statistically *efficient*, i.e., may gain *precision* using confidence interval

The most popular method for matching is called **category matching.** This involves first categorizing each of the matching factors and then finding, for each case, one or more controls from the same combined set of matching categories.

For example, if we are matching on age, race, and sex, we first categorize each of these three variables separately. For each case, we then determine his or her age–race–sex combination. For instance, the case may be 52 years old, white, and female. We then find one or more controls with the same age–race–sex combination.

If our study involves matching, we must decide on the number of controls to be chosen for each case. If we decide to use only one control for each case, we call this one-to-one or pair-matching. If we choose R controls for each case, for example, R equals 4, then we call this R-to-1 matching.

It is also possible to match so that there are different numbers of controls for different cases. That is, R may vary from case to case. For example, for some cases, there may be three controls, whereas for other cases perhaps only two or one control. This frequently happens when it is intended to do R-to-1 matching, but it is not always possible to find a full complement of R controls in the same matching category for some cases.

As for whether to match or not in a given study, there are both advantages and disadvantages to consider.

The primary advantage for matching over random sampling without matching is that matching can often lead to a more statistically efficient analysis. In particular, **matching may lead to a tighter confidence interval, that is, more precision,** around the odds or risk ratio being estimated than would be achieved without matching.

Disadvantage

    Matching is *costly*
- to find matches
- information loss due to discarding controls

The major disadvantage to matching is that it can be costly, both in terms of the time and labor required to find appropriate matches and in terms of information loss due to discarding of available controls not able to satisfy matching criteria. In fact, if too much information is lost from matching, it may be possible to lose statistical efficiency by matching.

Safest strategy

    Match on strong risk factors expected to be confounders

In deciding whether to match or not on a given factor, the safest strategy is to match only on strong risk factors expected to cause confounding in the data.

Matching	No matching
correct estimate?	
YES	YES
Apropriate analysis?	
YES	YES
↓	↓
MATCHED (STRATIFIED) ANALYSIS	STANDARD STRATIFIED ANALYSIS
↓	30–39 40–49 50–59
SEE SECTION III	$\widehat{OR}_1$ $\widehat{OR}_2$ $\widehat{OR}_3$
	combine

Note that whether one matches or not, it is possible to obtain an unbiased estimate of the effect, namely, the correct odds ratio estimate. The correct estimate can be obtained provided an appropriate analysis of the data is carried out.

If, for example, we match on age, the appropriate analysis is a **matched analysis,** which is a **special kind of stratified analysis** to be described shortly.

If, on the other hand, we do not match on age, an appropriate analysis involves dividing the data into age strata and doing a **standard stratified analysis** which combines the results from different age strata.

Validity is not an important reason for matching (validity: getting the right answer)

Because a correct estimate can be obtained whether or not one matches at the design stage, it follows that validity is not an important reason for matching. Validity concerns getting the right answer, which can be obtained by doing the appropriate stratified analysis.

Match to gain efficiency or precision

As mentioned above, the most important statistical reason for matching is to gain efficiency or precision in estimating the odds or risk ratio of interest. That is, matching becomes worthwhile if it leads to a tighter confidence interval than would be obtained by not matching.

III. Matched Analyses Using Stratification

Strata = matched sets

The analysis of matched data can be carried out using a stratified analysis in which the strata consist of the collection of matched sets.

Special case
  Case-control study
  100 matched pairs
  $n = 200$
  100 strata = 100 matched pairs
  2 observations per stratum

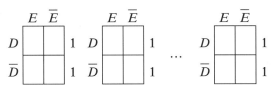

Four possible forms:

	E	$\overline{E}$	
D	1	0	1
$\overline{D}$	1	0	1

W pairs

	E	$\overline{E}$	
D	1	0	1
$\overline{D}$	0	1	1

X pairs

	E	$\overline{E}$	
D	0	1	1
$\overline{D}$	1	0	1

Y pairs

	E	$\overline{E}$	
D	0	1	1
$\overline{D}$	0	1	1

Z pairs

$W + X + Y + Z$ = total number of pairs

EXAMPLE

$W = 30, X = 30, Y = 10, Z = 30$
$W + X + Y + Z = 30 + 30 + 10 + 30 = 100$

Analysis: two equivalent ways

As a special case, consider a pair-matched case-control study involving 100 matched pairs. The total number of observations, n, then equals 200, and the data consists of 100 strata, each of which contains the two observations in a given matched pair.

If the only variables being controlled in the analysis are those involved in the matching, then the complete data set for this matched pairs study can be represented by 100 2×2 tables, one for each matched pair. Each table is labeled by exposure status on one axis and disease status on the other axis. The number of observations in each table is two, one being diseased and the other (representing the control) being nondiseased.

Depending on the exposure status results for this data, there are four possible forms that a given stratum can take. These are shown here.

The first of these contains a matched pair for which both the case and the control are exposed.

The second of these contains a matched pair for which the case is exposed and the control is unexposed.

In the third table, the case is unexposed and the control is exposed.

And in the fourth table, both the case and the control are unexposed.

If we let W, X, Y, and Z denote the number of pairs in each of the above four types of tables, respectively, then the sum W plus X plus Y plus Z equals 100, the total number of matched pairs in the study.

For example, we may have W equal 30, X equal 30, Y equal 10, and Z equal 30, which sums to 100.

The analysis of a matched pair data set can then proceed in either of two equivalent ways, which we now briefly describe.

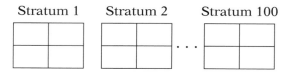

Stratum 1   Stratum 2   Stratum 100

Compute Mantel–Haenszel χ^2 and $\widehat{MOR}$

One way is to carry out a **Mantel–Haenszel chi-square test** for association based on the 100 strata and to compute a **Mantel–Haenszel odds ratio,** usually denoted as MOR, as a summary odds ratio that adjusts for the matched variables. This can be carried out using any standard computer program for stratified analysis. See Kleinbaum et al., *Epidemiologic Research,* (Von Nostrand Reinhold, 1982) for details.

The other method of analysis, which is equivalent to the above stratified analysis approach, is to summarize the data in a single table, as shown here. In this table, matched pairs are counted once, so that the total number of matched pairs is 100.

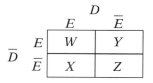

As described earlier, the quantity W represents the number of matched pairs in which both the case and the control are exposed. Similarly, X, Y, and Z are defined as previously.

$$\chi^2 = \frac{(X - Y)^2}{X + Y}, \quad df = 1$$

McNemar's test

Using the above table, the test for an overall effect of exposure, controlling for the matching variables can be carried out using a chi-square statistic equal to the square of the difference $X - Y$ divided by the sum of X and Y. This chi-square statistic has one degree of freedom in large samples and is called **McNemar's test.**

McNemar's test = MH test for pair-matching

$\widehat{MOR} = X/Y$

It can be shown that McNemar's test statistic is exactly equal to the Mantel–Haenszel chi-square statistic obtained by looking at the data in 100 strata. Moreover, the Mantel–Haenszel odds ratio estimate can be calculated as X/Y.

As an example of McNemar's test, suppose W equals 30, X equals 30, Y equals 10, and Z equals 30, as shown in the table here.

Then based on this data, the McNemar test statistic is computed as the square of 30 minus 10 divided by 30 plus 10, which equals 400 over 40, which equals 10.

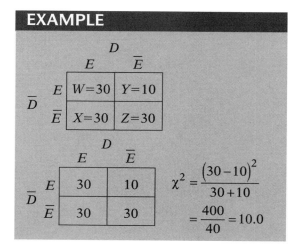

EXAMPLE

EXAMPLE (continued)

χ^2 ~chi square 1 df
under H_0: OR = 1

$P < <0.01$, significant

$\widehat{MOR} = \dfrac{X}{Y} = 3$

This statistic has approximately a chi-square distribution with 1 degree of freedom under the null hypothesis that the odds ratio relating exposure to disease equals 1.

From chi-square tables, we find this statistic to be highly significant with a P-value well below 0.01.

The estimated odds ratio, which adjusts for the matching variables, can be computed from the above table using the MOR formula X over Y, which in this case turns out to be 3.

Analysis for R-to-1 and mixed matching use stratified analysis

We have thus described how to do a matched pair analysis using stratified analysis or an equivalent McNemar's procedure. If the matching is R-to-1 or even involves mixed matching ratios, the analysis can also be done using a stratified analysis.

EXAMPLE

$R = 4$: Illustrating one stratum

	E	$\bar{E}$	
D	1	0	1
$\bar{D}$	1	3	4
			5

R-to-1 or mixed matching

use χ^2 (MH) and $\widehat{MOR}$ for stratified data

For example, if R equals 4, then each stratum contains five subjects, consisting of the one case and its four controls. These numbers can be seen on the margins of the table shown here. The numbers inside the table describe the numbers exposed and unexposed within each disease category. Here, we illustrate that the case is exposed, and that three of the four controls are unexposed. The breakdown within the table may differ with different matched sets.

Nevertheless, the analysis for R-to-1 or mixed matched data can proceed as with pair-matching by computing a Mantel–Haenszel chi-square statistic and a Mantel–Haenszel odds ratio estimate based on the stratified data.

IV. The Logistic Model for Matched Data
   1. Stratified analysis
   2. McNemar analysis
   ✓ 3. Logistic modeling

Advantage of modeling
   can control for variables *other* than matched variables

A third approach to carrying out the analysis of matched data involves logistic regression modeling.

The main advantage of using logistic regression with matched data occurs when there are variables other than the matched variables that the investigator wishes to control.

Match on AGE, RACE, SEX
also, control for SBP and BODYSIZE

For example, one may match on AGE, RACE, and SEX, but may also wish to control for systolic blood pressure and body size, which may have also been measured but were not part of the matching.

Logistic model for matched data includes control of variables not matched

In the remainder of the presentation, we describe how to formulate and apply a logistic model to analyze matched data, which allows for the control of variables not involved in the matching.

Stratified analysis inefficient:
  data is discarded

In this situation, using a stratified analysis approach instead of logistic regression will usually be inefficient in that much of one's data will need to be discarded, which is not required using a modeling approach.

Matched data:
  use conditional ML estimation
  (number of parameters large relative to n)

The model that we describe below for matched data requires the use of conditional ML estimation for estimating parameters. This is because, as we shall see, when there are matched data, the number of parameters in the model is large relative to the number of observations.

Pair-matching:

$$\widehat{OR}_U = (\widehat{OR}_C)^2$$
↑
overestimate

If unconditional ML estimation is used instead of conditional, an overestimate will be obtained. In particular, for pair-matching, the estimated odds ratio using the unconditional approach will be the square of the estimated odds ratio obtained from the conditional approach, the latter being the correct result.

Principle
  matched analysis ⇒ stratified analysis

- strata are matched sets, e.g., pairs
- strata defined using dummy (indicator) variables

An important principle about modeling matched data is that such modeling requires the matched data to be considered in strata. As described earlier, the strata are the matched sets, for example, the pairs in a matched pair design. In particular, the strata are defined using **dummy** or indicator variables, which we will illustrate shortly.

$E = (0, 1)$ exposure

$C_1, C_2, \ldots, C_p$ control variables

In defining a model for a matched analysis, we consider the special case of a single $(0, 1)$ exposure variable of primary interest, together with a collection of control variables C_1, C_2, and so on up through C_p, to be adjusted in the analysis for possible confounding and interaction effects.

- some C's matched by design
- remaining C's not matched

We assume that some of these C variables have been matched in the study design, either using pair-matching or R-to-1 matching. The remaining C variables have not been matched, but it is of interest to control for them, nevertheless.

$D = (0, 1)$ disease
$X_1 = E = (0, 1)$ exposure

Given the above context, we now define the following set of variables to be incorporated into a logistic model for matched data. We have a $(0, 1)$ disease variable D, and a $(0, 1)$ exposure variable X_1 equal to E.

Some X's: V_{1i} dummy variables (matched strata)

We also have a collection of X's which are dummy variables to indicate the different matched strata; these variables are denoted as V_1 variables.

Some X's: V_{2i} variables (potential confounders)

Further, we have a collection of X's, which are defined from the C's not involved in the matching and represent potential confounders in addition to the matched variables. These potential confounders are denoted as V_2 variables.

Some X's: product terms EW_j
(Note: W's usually V_2's)

And finally, we have a collection of X's which are product terms of the form E times W, where the W's denote potential interaction variables. Note that the W's will usually be defined in terms of the V_2 variables.

The model

$$\text{logit } P(\mathbf{X}) = \alpha + \beta E$$
$$+ \underbrace{\sum \gamma_{1i} V_{1i}}_{\text{matching}} + \underbrace{\sum \gamma_{2i} V_{2i}}_{\text{confounders}}$$
$$+ E \underbrace{\sum \delta_j W_j}_{\text{interaction}}$$

The logistic model for matched analysis is then given in logit form as shown here. In this model, the γ_{1i} are coefficients of the dummy variables for the matching strata, the γ_{2i} are the coefficients of the potential confounders not involved in the matching, and the δ_j are the coefficients of the interaction variables.

EXAMPLE

Pair-matching by AGE, RACE, SEX
100 matched pairs
99 dummy variables

$$V_{1i} = \begin{cases} 1 & \text{if } i\text{th matched pair} \\ 0 & \text{otherwise} \end{cases}$$
$$i = 1, 2, \ldots, 99$$

$$V_{11} = \begin{cases} 1 & \text{if first matched pair} \\ 0 & \text{otherwise} \end{cases}$$

As an example of dummy variables defined for matched strata, consider a study involving pair-matching by AGE, RACE, and SEX, containing 100 matched pairs. Then the above model requires defining 99 dummy variables to incorporate the 100 matched pairs.

We can define these dummy variables as V_{1i} equals 1 if an individual falls into the ith matched pair and 0 otherwise. Thus, it follows that

V_{11} equals 1 if an individual is in the first matched pair and 0 otherwise,

EXAMPLE (continued)

$$V_{12} = \begin{cases} 1 & \text{if second matched pair} \\ 0 & \text{otherwise} \end{cases}$$

$$\vdots$$

$$V_{1, 99} = \begin{cases} 1 & \text{if 99th matched pair} \\ 0 & \text{otherwise} \end{cases}$$

1st matched set
$$V_{11} = 1, V_{12} = V_{13} = \dots V_{1, 99} = 0$$
99th matched set
$$V_{1, 99} = 1, V_{11} = V_{12} = \dots V_{1, 98} = 0$$
100th matched set
$$V_{11} = V_{12} = \dots V_{1, 99} = 0$$

V_{12} equals 1 if an individual is in the second matched pair and 0 otherwise, and so on up to

$V_{1, 99}$, which equals 1 if an individual is in the 99th matched pair and 0 otherwise.

Alternatively, using the above dummy variable definition, a person in the first matched set will have V_{11} equal to 1 and the remaining dummy variables equal to 0; a person in the 99th matched set will have $V_{1, 99}$ equal to 1 and the other dummy variables equal to 0; and a person in the 100th matched set will have all 99 dummy variables equal to 0.

Matched pairs model
$$\text{logit } P(\mathbf{X}) = \alpha + \beta E + \sum \gamma_{1i} V_{1i} + \sum \gamma_{2i} V_{2i} + E \sum \delta_j W_j$$

For the matched analysis model, we have just described, the odds ratio formula for the effect of exposure status adjusted for covariates is given by the expression ROR equals e to the quantity β plus the sum of the δ_j times the W_j.

$$\text{ROR} = \exp\left(\beta + \sum \delta_j W_j\right)$$

Note: two types of V variables are controlled

This is exactly the same odds ratio formula given in our review for the E, V, W model. This makes sense because the matched analysis model is essentially an E, V, W model containing two different types of V variables.

V. An Application

EXAMPLE

Case-control study
2-to-1 matching

$D = \text{MI}_{0, 1}$
$E = \text{SMK}_{0, 1}$

$\underbrace{C_1 = \text{AGE}, C_2 = \text{RACE}, C_3 = \text{SEX}, C_4 = \text{HOSPITAL}}_{\text{matched}}$

$\underbrace{C_5 = \text{SBP} \quad C_6 = \text{ECG}}_{\text{not matched}}$

As an application of a matched pairs analysis, consider a case-control study involving 2-to-1 matching which involves the following variables:

The **disease variable** is myocardial infarction status, as denoted by MI.

The **exposure variable** is smoking status, as defined by a (0, 1) variable denoted as SMK.

There are six C variables to be controlled. The first four of these variables, namely, age, race, sex, and hospital status, are involved in the matching.

The last two variables, systolic blood pressure, denoted by SBP, and electrocardiogram status, denoted by ECG, are not involved in the matching.

EXAMPLE (continued)

$n = 117$ (39 matched sets)

The model:

$$\text{logit P}(\mathbf{X}) = \alpha + \beta\text{SMK} + \sum_{i=1}^{38} \gamma_{1i} V_{1i}$$

$$= \gamma_{21}\ \text{SBP} + \gamma_{22}\ \text{ECG}$$
$$\underbrace{\qquad\qquad}_{\text{confounders}}$$
$$+\text{SMK}\big(\delta_1\text{SBP} + \delta_2\text{ECG}\big)$$
$$\underbrace{\qquad\qquad}_{\text{modifiers}}$$

$$\text{ROR} = \exp\big(\beta + \delta_1\text{SBP} + \delta_2\text{ECG}\big)$$

$\beta = $ coefficient of E

$\delta_1 = $ coefficient of $E \times \text{SBP}$

$\delta_2 = $ coefficient of $E \times \text{ECG}$

Starting model

analysis strategy

Final model

Estimation method:
   ✓ conditional ML estimation
   (also, we illustrate unconditional ML estimation)

Interaction:
SMK $\times$ SBP and SMK $\times$ ECG?

The study involves 117 persons in 39 matched sets, or strata, each strata containing 3 persons, 1 of whom is a case and the other 2 are matched controls.

The logistic model for the above situation can be defined as follows: logit P($\mathbf{X}$) equals α plus β times SMK plus the sum of 38 terms of the form γ_{1i} times V_{1i}, where V_{1i} are dummy variables for the 39 matched sets, plus γ_{21} times SBP plus γ_{22} times ECG plus SMK times the sum of δ_1 SBP plus δ_2 times ECG.

Here, we are considering two potential confounders involving the two variables (SBP and ECG) not involved in the matching and also two interaction variables involving these same two variables.

The odds ratio for the above logistic model is given by the formula e to the quantity β plus the sum of δ_1 times SBP and δ_2 times ECG.

Note that this odds ratio expression involves the coefficients β, δ_1, and δ_2, which are coefficients of variables involving the exposure variable. In particular, δ_1 and δ_2 are coefficients of the interaction terms E $\times$ SBP and E $\times$ ECG.

The model we have just described is the starting model for the analysis of the data set on 117 subjects. We now address how to carry out an analysis strategy for obtaining a final model that includes only the most relevant of the covariates being considered initially.

The first important issue in the analysis concerns the choice of estimation method for obtaining ML estimates. Because matching is being used, the appropriate method is conditional ML estimation. Nevertheless, we also show the results of unconditional ML estimation to illustrate the type of bias that can result from using the wrong estimation method.

The next issue to be considered is the assessment of interaction. Based on our starting model, we, therefore, determine whether or not either or both of the product terms SMK $\times$ SBP and SMK $\times$ ECG are retained in the model.

EXAMPLE (continued)

Chunk test

$$H_0: \delta_1 = \delta_2 = 0$$
where
δ_1 = coefficient of SMK $\times$ SBP
δ_2 = coefficient of SMK $\times$ ECG

$$LR = \left(-2 \ln \hat{L}_R\right) - \left(-2 \ln \hat{L}_F\right)$$

R = reduced model	F = full model
(no interaction)	(interaction)

log likelihood statistics
$$-2 \ln \hat{L}$$

$$LR \sim \chi^2_2$$
Number of parameters tested = 2

$$-2 \ln \hat{L}_F = 60.23$$
$$-2 \ln \hat{L}_R = 60.63$$

$$LR = 60.63 - 60.23 = 0.40$$

$P > 0.10$ (no significant interaction)

Therefore, drop SMK $\times$ SBP and SMK $\times$ ECG from model

Backward elimination: same conclusion

$$\text{logit } P(\mathbf{X}) = \alpha + \beta SMK + \sum \gamma_{1i} V_{1i}$$
$$+ \gamma_{21} SBP + \gamma_{22} ECG$$

One way to test for this interaction is to carry out a chunk test for the significance of both product terms considered collectively. This involves testing the null hypothesis that the coefficients of these variables, namely, δ_1 and δ_2, are both equal to 0.

The test statistic for this chunk test is given by the likelihood ratio (LR) statistic computed as the difference between log likelihood statistics for the full model containing both interaction terms and a reduced model which excludes both interaction terms. The log likelihood statistics are of the form $-2 \ln L$ "hat," where L "hat" is the maximized likelihood for a given model.

This likelihood ratio statistic has a chi-square distribution with two degrees of freedom. The degrees of freedom are the number of parameters tested, namely, 2.

When carrying out this test, the log likelihood statistics for the full and reduced models turn out to be 60.23 and 60.63, respectively.

The difference between these statistics is 0.40. Using chi-square tables with two degrees of freedom, the P-value is considerably larger than 0.10, so we can conclude that there are no significant interaction effects. We can, therefore, drop the two interaction terms from the model.

Note that an alternative approach to testing for interaction is to use backward elimination on the interaction terms in the initial model. Using this latter approach, it turns out that both interaction terms are eliminated. This strengthens the conclusion of no interaction.

At this point, our model can be simplified to the one shown here, which contains only main effect terms. This model contains the exposure variable SMK, 38 V variables that incorporate the 39 matching strata, and 2 V variables that consider the potential confounding effects of SBP and ECG, respectively.

EXAMPLE (continued)

$$\widehat{ROR} = e^{\hat{\beta}}$$

V's in model	OR = e^{β}	95% CI
SBP and ECG	C (2.14)	(0.70, 6.50)
	U 3.53	
SBP only	C 2.08	(0.72, 6.00)
	U 3.39	
ECG only	C 2.05	(0.74, 5.70)
	U 3.07	
Neither	C 2.32	(0.93, 5.79)
	U 3.71	

C = conditional estimate
U = unconditional estimate

Minimal confounding:
   gold standard $\widehat{OR}$ = 2.14, essentially
   same as other $\widehat{OR}$

But 2.14 moderately different from 2.32, so we control for *at least* one of SBP and ECG

Narrowest CI: control for ECG only

Most precise estimate:
   control for ECG only

All CI are wide and include 1

Overall conclusion:
   adjusted $\widehat{OR}$ ≈ 2, but is
   nonsignificant

Under this reduced model, the estimated odds ratio adjusted for the effects of the V variables is given by the familiar expression e to the β "hat," where β "hat" is the coefficient of the exposure variable SMK.

The results from fitting this model and reduced versions of this model which delete either or both of the potential confounders SBP and ECG are shown here. These results give both conditional (C) and unconditional (U) odds ratio estimates and 95% confidence intervals for the conditional estimates only.

From inspection of this table of results, we see that the unconditional estimation procedure leads to overestimation of the odds ratio and, therefore, should not be used.

The results also indicate a minimal amount of confounding due to SBP and ECG. This can be seen by noting that the gold standard estimated odds ratio of 2.14, which controls for both SBP and ECG, is essentially the same as the other conditionally estimated odds ratios that control for either SBP or ECG or neither.

Nevertheless, because the estimated odds ratio of 2.32, which ignores both SBP and ECG in the model, is moderately different from 2.14, we recommend that at least one or possibly both of these variables be controlled.

If at least one of SBP and ECG is controlled, and confidence intervals are compared, the narrowest confidence interval is obtained when only ECG is controlled.

Thus, the most precise estimate of the effect is obtained when ECG is controlled, along, of course, with the matching variables.

Nevertheless, because all confidence intervals are quite wide and include the null value of 1, it does not really matter which variables are controlled. The overall conclusion from this analysis is that the adjusted estimate of the odds ratio for the effect of smoking on the development of MI is about 2, but it is quite nonsignificant.

SUMMARY

This presentation
- basic features of matching
- logistic model for matched data
- illustration using 2-to-1 matching

This presentation is now complete. In summary, we have described the basic features of matching, presented a logistic regression model for the analysis of matched data, and have illustrated the model using an example from a 2-to-1 matched case-control study.

Logistic Regression Chapters
1. Introduction
2. Important Special Cases
   .
   .
   .
✓ (8. Analysis of Matched Data)

The reader may wish review the detailed summary and to try the practice exercises and the test that follow.

This is the last presentation in this text on logistic regression. For current applications of logistic regression, the reader may consider published work in popular epidemiologic journals such as the *American Journal of Epidemiologic Research* and the *International Journal of Epidemiology.*

**Detailed
Outline**

I. **Overview** (page 230)

Focus

- basics of matching
- model for matched data
- control of confounding and interaction
- examples

II. **Basic features of matching** (pages 230–232)

  A. Study design procedure: select referent group to be constrained so as to be comparable to index group on one or more factors:

   i. case-control study (our focus): referent=controls, index=cases;

   ii. follow-up study: referent=unexposed, index=exposed.

  B. Category matching: if case-control study, find, for each case, one or more controls in the same combined set of categories of matching factors.

  C. Types of matching: 1-to-1, R-to-1, other.

  D. To match or not to match:

   i. advantage: can gain efficiency/precision;

   ii. disadvantages: costly to find matches and might lose information discarding controls;

   iii. safest strategy: match on strong risk factors expected to be confounders;

   iv. validity not a reason for matching: can get valid answer even when not matching.

III. **Matched analyses using stratification** (pages 232–235)

  A. Strata are matched sets, e.g., if 4-to-1 matching, each stratum contains five observations.

  B. Special case: 1-to-1 matching: four possible forms of strata:

   i. both case and control are exposed (W pairs);

   ii. only case is exposed (X pairs);

   iii. only control is exposed (Y pairs),

   iv. neither case nor control is exposed (Z pairs).

  C. Two equivalent analysis procedures for 1-to-1 matching:

   i. Mantel–Haenszel (MH): use MH test on all strata and compute MOR estimate of OR;

   ii. McNemar approach: group data by pairs (W, X, Y, and Z as in B above). Use McNemar's chi-square statistic $(X-Y)^2/(X+Y)$ for test and X/Y for estimate of OR.

  D. R-to-1 matching: use MH test statistic and MOR.

IV. The logistic model for matched data (pages 235–238)

    A. Advantage: provides an efficient analysis when there are variables other than matching variables to control.

    B. Model uses dummy variables in identifying different strata.

    C. Model form:

$$\text{logit } P(\mathbf{X}) = \alpha + \beta E + \sum \gamma_{1i} V_{1i} + \sum \gamma_{2i} V_{2i} + E \sum \delta_j W_j$$

    where V_{1i} are dummy variables identifying matched strata, V_{2i} are potential confounders based on variables not involved in the matching, and W_j's are effect modifiers (usually) based on variables not involved in the matching.

    D. Odds ratio expression if E is coded as (0, 1):

$$\text{ROR} = \exp\left(\beta + \sum \delta_j W_j\right)$$

V. An application (pages 238–241)

    A. Case-control study, 2-to-1 matching, D = MI (0, 1),
    E = SMK (0, 1),
    four matching variables: AGE, RACE, SEX, HOSPITAL,
    two variables not matched: SBP, ECG,
    n = 117 (39 matched sets, 3 observations per set).

    B. Model form:

$$\text{logit } P(\mathbf{X}) = \alpha + \beta \text{SMK} + \sum_{i=1}^{38} \gamma_{1i} V_{1i} + \gamma_{21}\text{SBP} + \gamma_{22}\text{ECG} + \text{SMK}(\delta_1\text{SBP} + \delta_2\text{ECG}).$$

    C. Odds ratio:

$$\text{ROR} = \exp\left(\beta + \delta_1\text{SBP} + \delta_2\text{ECG}\right).$$

    D. Analysis: use conditional ML estimation; interaction not significant;
    No interaction model:

$$\text{logit } P(\mathbf{X}) = \alpha + \beta \text{SMK} + \sum_{i=1}^{38} \gamma_{1i} V_{1i} + \gamma_{21}\text{SBP} + \gamma_{22}\text{ECG}.$$

    Odds ratio formula:
    $\text{ROR} = \exp(\beta)$,
    Gold standard OR estimate controlling for SBP and ECG: 2.14,
    Narrowest CI obtained when only ECG is controlled: OR estimate is 2.08,
    Overall conclusion: OR approximately 2, but not significant.

Practice Exercises

True or False (Circle T or F)

T F 1. In a case-control study, category pair-matching on age and sex is a procedure by which, for each control in the study, a case is found as its pair to be in the same age category and same sex category as the control.

T F 2. In a follow-up study, pair-matching on age is a procedure by which the age distribution of cases (i.e., those with the disease) in the study is constrained to be the same as the age distribution of noncases in the study.

T F 3. In a 3-to-1 matched case-control study, the number of observations in each stratum, assuming sufficient controls are found for each case, is four.

T F 4. An advantage of matching over not matching is that a more precise estimate of the odds ratio may be obtained from matching.

T F 5. One reason for deciding to match is to gain validity in estimating the odds ratio of interest.

T F 6. When in doubt, it is safer to match than not to match.

T F 7. A matched analysis can be carried out using a stratified analysis in which the strata consists of the collection of matched sets.

T F 8. In a pair-matched case-control study, the Mantel–Haenszel odds ratio (i.e., the MOR) is equivalent to McNemar's test statistic $(X - Y)^2/(X+Y)$. (Note: X denotes the number of pairs for which the case is exposed and the control is unexposed, and Y denotes the number of pairs for which the case is unexposed and the control is exposed.)

T F 9. When carrying out a Mantel–Haenszel chi-square test for 4-to-1 matched case-control data, the number of strata is equal to five.

T F 10. Suppose in a pair-matched case-control study, the number of pairs in each of the four cells of the table used for McNemar's test is given by $W = 50$, $X = 40$, $Y = 20$, and $Z = 100$. Then the computed value of McNemar's test statistic is given by 2.

11. For the pair-matched case-control study described in Exercise 10, let E denote the (0, 1) exposure variable and let D denote the (0, 1) disease variable. State the logit form of the logistic model that can be used to analyze this data. (Note: Other than the variables matched, there are no other control variables to be considered here.)

12. Consider again the pair-matched case-control data described in Exercise 10 ($W = 50, X = 40, Y = 20, Z = 100$). Using conditional ML estimation, a logistic model fitted to this data resulted in an estimated coefficient of exposure equal to 0.693, with standard error equal to 0.274. Using this information, compute an estimate of the odds ratio of interest and compare its value with the estimate obtained using the MOR formula X/Y.

13. For the same situation as in Exercise 12, compute the Wald test for the significance of the exposure variable and compare its squared value and test conclusion with that obtained using McNemar's test.

14. Use the information provided in Exercise 12 to compute a 95% confidence interval for the odds ratio, and interpret your result.

15. If unconditional ML estimation had been used instead of conditional ML estimation, what estimate would have been obtained for the odds ratio of interest? Which estimation method is correct, conditional or unconditional, for this data set?

Consider a 2-to-1 matched case-control study involving 300 bisexual males, 100 of whom are cases with positive HIV status, with the remaining 200 being HIV negative. The matching variables are AGE and RACE. Also, the following additional variables are to be controlled but are not involved in the matching: NP, the number of sexual partners within the past 3 years; ASCM, the average number of sexual contacts per month over the past 3 years, and PAR, a (0, 1) variable indicating whether or not any sexual partners in the past 5 years were in high-risk groups for HIV infection. The exposure variable is CON, a (0, 1) variable indicating whether the subject used consistent and correct condom use during the past 5 years.

16. Based on the above scenario, state the logit form of a logistic model for assessing the effect of CON on HIV acquisition, controlling for NP, ASCM, and PAR as potential confounders and PAR as the only effect modifier.

17. Using the model given in Exercise 16, give an expression for the odds ratio for the effect of CON on HIV status, controlling for the confounding effects of AGE, RACE, NP, ASCM, and PAR, and for the interaction effect of PAR.

18. For the model used in Exercise 16, describe the strategy you would use to arrive at a final model that controls for confounding and interaction.

Test

True or False (Circle T or F)

T F 1. In a category matched 2-to-1 case-control study, each case is matched to two controls who are in the same category as the case for each of the matching factors.

T F 2. An advantage of matching over not matching is that information may be lost when not matching.

T F 3. If we do not match on an important risk factor for the disease, it is still possible to obtain an unbiased estimate of the odds ratio by doing an appropriate analysis that controls for the important risk factor.

T F 4. McNemar's test statistic is not appropriate when there is R-to-1 matching and R is at least 2.

T F 5. In a matched case-control study, logistic regression can be used when it is desired to control for variables involved in the matching as well as variables not involved in the matching.

6. Consider the following McNemar's table from the "Agent Orange" study analyzed by Donovan et al. (*Med. J. of Australia*, 1984). This is a pair-matched case-control study, where the cases are babies born with genetic anomalies and controls are babies born without such anomalies. The matching variables are hospital, time period of birth, mother's age, and health insurance status. The exposure factor is status of father (Vietnam veteran = 1 or nonveteran = 0):

		Case	
		E	not E
Control	E	2	121
	not E	125	8254

For the above data, carry out McNemar's test for the significance of exposure and compute the estimated odds ratio. What are your conclusions?

7. State the logit form of the logistic model that can be used to analyze the Agent Orange study data.

8. The following printout results from using conditional ML estimation of an appropriate logistic model for analyzing the Agent Orange data:

Variable	β	s_β	P-value	OR	95% CI for OR L	U
E	0.032	0.128	0.901	1.033	0.804	1.326

Use these results to compute the squared Wald test statistic for testing the significance of exposure and compare this test statistic with the McNemar chi-square statistic computed in Question 6.

9. How does the odds ratio obtained from the printout given in Question 8 compare with the odds ratio computed using McNemar's formula X/Y?

10. Explain how the confidence interval given in the printout is computed.

The following questions consider information obtained from a matched case-control study of cervical cancer in 313 women from Sydney, Australia (Brock et al., *J. Nat. Cancer Inst.*, 1988). The outcome variable is cervical cancer status (1 = present, 0 = absent). The matching variables are age and socioeconomic status. Additional independent variables not matched on are smoking status, number of lifetime sexual partners, and age at first sexual intercourse. The independent variables not involved in the matching are listed below together with their computer abbreviation and coding scheme.

Variable	Abbreviation	Coding
Smoking status	SMK	1 = ever, 0 = never
Number of sexual partners	NS	1 = 4+, 0 = 0 − 3
Age at first intercourse	AS	1 = 20+, 0 = ≤19

PRINTOUT:

Variable	β	S.E.	Chi sq	P
SMK	1.9381	0.4312	20.20	0.0000
NS	1.4963	0.4372	11.71	0.0006
AS	−0.6811	0.3473	3.85	0.0499
SMK × NS	−1.1128	0.5997	3.44	0.0635

11. What method of estimation was used to obtain estimates given in the above printout? Explain.

12. Why are the variables age and socioeconomic status missing from the printout given above, even though these were variables matched on in the study design?

13. State the logit form of the model used in the above printout.

14. Based on the printout above, is the product term SMK×NS significant? Explain.

15. Using the printout, give a formula for the point estimate of the odds ratio for the effect of SMK on cervical cancer status which adjusts for the confounding effects of NS and AS and allows for the interaction of NS with SMK.

16. Use the formula computed in Question 15 to compute numerical values for the estimated odds ratios when NS = 1 and NS = 0.

17. When NS = 1, the 95% confidence interval for the adjusted odds ratio for the effect of smoking on cervical cancer status is given by the limits (0.96, 5.44). Use this result and your estimate from Question 16 for NS = 1 to draw conclusions about the effect of smoking on cervical cancer status when NS = 1.

18. The following printout results from fitting a no interaction model to the cervical cancer data:

Variable	β	S.E.	Chi sq	P
SMK	1.4361	0.3167	20.56	0.0000
NS	0.9598	0.3057	9.86	0.0017
AS	-0.6064	0.3341	3.29	0.0695

Based on this printout, compute the odds ratio for the effect of smoking, test its significance, and derive a 95% confidence interval of the odds ratio. Based on these results, what do you conclude about the effect of smoking on cervical cancer status?

Answers to Practice Exercises

1. F: cases are selected first, and controls are matched to cases
2. F: the age distribution for exposed persons is constrained to be the same as for unexposed persons
3. T
4. T
5. F: matching is not needed to obtain a valid estimate of effect
6. F: when in doubt, matching may not lead to increased precision; it is safe to match only if the potential matching factors are strong risk factors expected to be confounders in the data
7. T
8. F: the Mantel–Haenszel chi-square statistic is equal to McNemar's test statistic
9. F: the number of strata equals the number of matched sets
10. F: the computed value of McNemar's test statistic is 6.67; the MOR is 2
11.

$$\operatorname{logit} P(\mathbf{X}) = \alpha + \beta E + \sum_{i=1}^{209} \gamma_{1i} V_{1i},$$

where the V_{1i} denote dummy variables indicating the different matched pairs (strata).

12. Using the printout, the estimated odds ratio is exp(0.693), which equals 1.9997. The MOR is computed as X/Y equals 40/20 equals 2. Thus, the estimate obtained using conditional logistic regression is equal to the MOR.

13. The Wald statistic, which is a Z statistic, is computed as 0.693/0.274, which equals 2.5292. This is significant at the 0.01 level of significance, i.e., P is less than 0.01. The squared Wald statistic, which has a chi-square distribution with one degree of freedom under the null hypothesis of no effect, is computed to be 6.40. The McNemar chi-square statistic is 6.67, which is quite similar to the Wald result, though not exactly the same.

14. The 95% confidence interval for the odds ratio is given by the formula

$$\exp\left[\hat{\beta} \pm 1.96\sqrt{\widehat{\text{var}}(\hat{\beta})}\right]$$

which is computed to be

$\exp(0.693 \pm 1.96 \times 0.274) = \exp(0.693 \pm 0.53704)$

which equals $(e^{0.15596}, e^{1.23004}) = (1.17, 3.42)$.

This confidence interval around the point estimate of 2 indicates that the point estimate is somewhat unstable. In particular, the lower limit is close to the null value of 1, whereas the upper limit is close to 4. Note also that the confidence interval does not include the null value, which supports the statistical significance found in Exercise 13.

15. If unconditional ML estimation had been used, the odds ratio estimate would be higher (i.e., an overestimate) than the estimate obtained using conditional ML estimation. In particular, because the study involved pair-matching, the unconditional odds ratio is the square of the conditional odds ratio estimate. Thus, for this data set, the conditional estimate is given by MOR equal to 2, whereas the unconditional estimate is given by the square of 2, or 4. The correct estimate is 2, not 4.

16.
$$\text{logit P}(\mathbf{X}) = \alpha + \beta\text{CON} + \sum_{i=1}^{99}\gamma_{1i}V_{1i} + \gamma_{21}\text{NP} + \gamma_{22}\text{ASCM} + \gamma_{23}\text{PAR} + \delta\text{CON} \times \text{PAR},$$

where the V_{1i} are 99 dummy variables indicating the 100 matching strata, with each stratum containing three observations.

17. $\widehat{\text{ROR}} = \exp(\hat{\beta} + \hat{\delta}\text{PAR})$.

18. A recommended strategy for model building involves first testing for the significance of the interaction term in the starting model given in Exercise 16. If this test is significant, then the final model must contain the interaction term, the main effect of PAR (from the Hierarchy Principle), and the 99 dummy variables for matching. The other two variables NP and ASCM may be dropped as nonconfounders if the odds ratio given by Exercise 17 does not meaningfully change when either or both variables are removed from the model. If the interaction test is not significant, then the reduced (no interaction) model is given by the expression

$$\text{logit } P(\mathbf{X}) = \alpha + \beta CON + \sum_{i=1}^{99} \gamma_{1i}V_{1i} + \gamma_{21}NP + \gamma_{22}ASCM + \gamma_{23}PAR.$$

Using this reduced model, the odds ratio formula is given by $\exp(\beta)$, where β is the coefficient of the CON variable. The final model must contain the 99 dummy variables which incorporate the matching into the model. However, NP, ASCM, and/or PAR may be dropped as nonconfounders if the odds ratio $\exp(\beta)$ does not change when one or more of these three variables are dropped from the model. Finally, precision of the estimate needs to be considered by comparing confidence intervals for the odds ratio. If a meaningful gain of precision is made by dropping a nonconfounder, then such a nonconfounder may be dropped. Otherwise (i.e., no gain in precision), the nonconfounder should remain in the model with all other variables needed for controlling confounding.

Appendix: Computer Data Sets

In this appendix, we provide listings of data sets that the reader or a course instructor may wish to use for practice exercises involving fitting logistic models using the computer. Sample questions for each data set are provided following each listing.

Data Set 1

Data set name: evans.dat

Variable 1 = number of coronary heart disease cases—CHD
2 = total number at risk in stratum
3 = serum catecholamine level (0 = low, 1 = high)—CAT
4 = age group (0 = below 55, 1 = 55+)—AGE
5 = electrocardiogram abnormality (0 = normal, 1 = abnormal)—ECG

17	274	0	0	0
15	122	0	1	0
7	59	0	0	1
5	32	0	1	1
1	8	1	0	0
9	39	1	1	0
3	17	1	0	1
14	58	1	1	1

Sample Questions

A. Fit a logistic model that predicts CHD from the predictors CAT, AGE, and ECG.

B. Fit a logistic model that predicts CHD from the predictors, CAT, AGE, ECG, CAT × AGE, and CAT × ECG.

C. Determine the estimated odds ratio for the effect of CAT adjusted for AGE and ECG based on the model fit in Question A.

D. Determine the formula for the estimated odds ratio for the effect of CAT adjusted for AGE and ECG based on the model fit in Question B. Use this odds ratio formula to derive individual odds ratio estimates for various choices of AGE and ECG.

E. Fit the same model as used in Question A, but code the CAT variable as −1=low, +1=high. Compare your results in Questions A and C with the corresponding results using the new coding.

F. Test for interaction using a likelihood ratio test.

G. Test for the significance of CAT adjusting for AGE and ECG using both Wald and LR tests.

H. Obtain 95% confidence intervals for the effect of CAT adjusted for AGE and ECG.

I. Use a hierarchical backward elimination strategy to obtain the "best" model.

Data Set 2

Data set name: bronchitis.dat
Variable 1 = number of bronchitis cases—BRC
         2 = total number at risk in stratum
         3 = smoking status (0 = never, 1 = ever)—SMK
         4 = socioeconomic status (0 = low, 1 = high)—SES
         5 = age (0 = below 40, 1 = 40–59)—AGE

38	111	0	1	0
48	134	0	1	1
28	95	0	0	0
40	124	0	0	1
84	173	1	1	0
102	148	1	1	1
47	143	1	0	0
59	112	1	0	1

Sample Questions

A. Fit a logistic model that predicts BRC from the predictors SMK, SES, and AGE.

B. Fit a logistic model that predicts BRC from the predictors SMK, SES, AGE, SMK $\times$ SES, and SMK $\times$ AGE.

C. Determine the estimated odds ratio for the effect of SMK adjusted for SES and AGE based on the model fit in Question A.

D. Determine the formula for the estimated odds ratio for the effect of SMK adjusted for SES and AGE based on the model fit in Question B. Use this odds ratio formula to derive individual odds ratio estimates for various choices of SES and AGE.

E. Fit the same model as used in Question A but code the SMK variables as -1 = low, $+1$ = high. Compare your results in Questions A and C with the corresponding results using the new coding.

F. Test for interaction using a likelihood ratio test.

G. Test for the significance of SMK adjusting for SES and AGE using both Wald and LR tests.

H. Obtain 95% confidence intervals for the effect of SMK adjusted for SES and AGE.

I. Use a hierarchical backward elimination strategy to obtain the "best" model.

Data Set 3

Data set name : conf.dat

Variable 1 = person id #
   2 = social class status—SOC
   3 = number of times victimized—VICT
   4 = AGE
   5 = index of confidence in legal system—CONF

1	low	0	4	25
2	low	0	14	45
3	low	0	15	46
4	low	0	19	25
5	low	0	17	48
6	low	0	16	30
7	med	0	7	39
8	med	0	10	35
9	med	0	12	39
10	med	0	15	43
11	med	0	16	44
12	high	0	8	37
13	high	0	19	40
14	high	0	10	40
15	high	0	17	25
16	low	1	2	24
17	low	1	7	25
18	low	1	18	46
19	med	1	6	40
20	med	1	19	35
21	med	1	12	43
22	med	1	12	45
23	high	1	7	33
24	high	1	6	32
25	high	1	5	39
26	high	1	3	30
27	high	1	16	31
28	low	2	7	36
29	low	2	8	38

30	low	2	2	36
31	low	2	11	41
32	low	2	12	40
33	med	2	1	28
34	med	2	2	35
35	med	2	4	27
36	high	2	4	27
37	high	2	2	37
38	high	2	8	35
39	high	2	11	41

Sample Questions

A.   Define a new variable that dichotomizes the CONF variable into high low groups, where high denotes a CONF value equal to or greater than 10 and low denotes a CONF value below 10. Also, define new variables from SOC and VICT as follows: SOC1 = 1 if low, 2 if medium, 3 if high; VICT1 = 0 if VICT is 0 or 1; and VICT1 = 1 if VICT is 2 or more.

B.   Carry out a logistic regression analysis involving the dichotomized CONF variable as the dependent variable with AGE, SOC1, and VICT1 as independent variables. Assume that the goal of the analysis is to assess the effect of the number of times victimized on confidence in the legal system, controlling for the possible confounding and effect modifying effects of age and social class. In your analysis, make sure to:

   i.   summarize the strategy used to obtain the best model.

   ii.   carry out an assessment of collinearity and remedy any collinearity problem that might be found.

  iii.   provide a summary of relevant computer results for the models you have run.

  iv.   provide estimated odds ratios for the effect of the number of times victimized on confidence in the legal system, controlling for age and social class, based on best model. Also, evaluate the significance of the odds ratios obtained and provide 95% confidence intervals for your odds ratio(s).

   v.   What conclusions do you draw from your analysis?

  vi.   How might you criticize the use of the variables SOC1 and VICT1 in this analysis? What alternatives would you suggest?

Data Set 4

Data set name: agentor.dat

These are data from a pair-matched case-control study, where the cases are babies born with genetic anomalies and controls are babies born without such anomalies. The matching variables are hospital, time period of birth, mother's age, and health insurance status. The exposure factor is status of father (Vietnam veteran = 1, nonveteran = 0):

		Case	
		E	not E
Control	E	2	121
	not E	125	8254

Sample Questions

A. Use the data in the above table to set up a data file that allows a conditional ML approach for fitting a logistic regression model that assesses the effect of exposure on case-control status.

B. Carry out conditional ML estimation to analyze these data. Use your results to obtain an odds ratio for the effect of exposure, test for the significance of this odds ratio, and obtain a 95% confidence interval for this odds ratio.

C. Compare your results from using conditional logistic regression to results obtained using a stratified (matched) analysis.

D. Use unconditional ML estimation to fit a logistic model to these data and compare your results to those obtained using conditional ML estimation.

Data Set 5

This data set is discussed in Chapter 8. The study involves 117 persons in 39 matched sets, each strata containing 3 persons, 1 of whom is a case and the other 2 are matched controls. The disease variable is myocardial infarction status (MI). The exposure variable is smoking status (SMK). Four variables are involved in the matching, namely, age, race, sex, and hospital status. Two additional independent variables, systolic blood pressure (SBP) and electrocardiogram status (ECG), are not involved in the matching.

Data set name: my.dat

Matched set	Person	Case status	SMK	SBP	ECG
1	1	1	0	160	1
1	2	0	0	140	0
1	3	0	0	120	0
2	4	1	0	160	1
2	5	0	0	140	0
2	6	0	0	120	0
3	7	1	0	160	0
3	8	0	0	140	0
3	9	0	0	120	0
4	10	1	0	160	0
4	11	0	0	140	0
4	12	0	0	120	0
5	13	1	0	160	0
5	14	0	0	140	0
5	15	0	0	120	0
6	16	1	0	160	0
6	17	0	0	140	0
6	18	0	0	120	0
7	19	1	0	160	0
7	20	0	0	140	0
7	21	0	0	120	0
8	22	1	0	160	0
8	23	0	0	140	0
8	24	0	0	120	0
9	25	1	0	160	0
9	26	0	0	140	0
9	27	0	0	120	0
10	28	1	0	160	0

Matched set	Person	Case status	SMK	SBP	ECG
10	29	0	0	140	0
10	30	0	0	120	0
11	31	1	0	120	1
11	32	0	0	120	0
11	33	0	0	120	0
12	34	1	0	120	0
12	35	0	0	120	0
12	36	0	0	120	0
13	37	1	0	120	0
13	38	0	0	120	0
13	39	0	0	120	0
14	40	1	0	140	0
14	41	0	0	140	0
14	42	0	0	140	0
15	43	1	0	120	1
15	44	0	0	140	1
15	45	0	0	160	0
16	46	1	0	120	1
16	47	0	0	140	1
16	48	0	0	160	1
17	49	1	1	160	1
17	50	0	0	140	0
17	51	0	0	120	0
18	52	1	1	160	1
18	53	0	0	140	0
18	54	0	0	120	0
19	55	1	1	160	0
19	56	0	0	140	1
19	57	0	0	120	0
20	58	1	1	160	1
20	59	0	0	140	1
20	60	0	0	120	0
21	61	1	1	160	0
21	62	0	0	140	0
21	63	0	0	120	0
22	64	1	1	120	0
22	65	0	0	120	0

Matched set	Person	Case status	SMK	SBP	ECG
22	66	0	0	120	0
23	67	1	1	140	0
23	68	0	0	140	0
23	69	0	0	140	0
24	70	1	1	120	0
24	71	0	0	140	0
24	72	0	0	160	0
25	73	1	1	120	0
25	74	0	0	160	0
25	75	0	0	140	0
26	76	1	0	160	0
26	77	0	1	140	0
26	78	0	0	120	0
27	79	1	0	120	0
27	80	0	1	120	0
27	81	0	0	120	0
28	82	1	0	160	1
28	83	0	0	140	0
28	84	0	1	120	0
29	85	1	0	160	0
29	86	0	0	140	0
29	87	0	1	120	0
30	88	1	0	120	0
30	89	0	0	140	0
30	90	0	1	160	0
31	91	1	0	140	0
31	92	0	0	140	0
31	93	0	1	140	0
32	94	1	1	160	1
32	95	0	1	140	0
32	96	0	0	120	0
33	97	1	1	160	1
33	98	0	1	140	1
33	99	0	0	120	0
34	100	1	1	120	1
34	101	0	1	120	1
34	102	0	0	120	1

Matched set	Person	Case status	SMK	SBP	ECG
35	103	1	1	160	0
35	104	0	0	140	0
35	105	0	1	120	0
36	106	1	0	160	1
36	107	0	1	140	1
36	108	0	1	120	1
37	109	1	0	120	0
37	110	0	1	140	0
37	111	0	1	160	0
38	112	1	1	160	1
38	113	0	1	140	0
38	114	0	1	120	0
39	115	1	1	120	0
39	116	0	1	120	0
39	117	0	1	120	0

Sample Questions

A. Set up a data file than can be used to carry out conditional ML estimation of a logistic model for these data.

B. Use conditional ML estimation to fit a model that allows for both confounding and interaction involving the variables SBP and ECG. Test for the significance of the interaction terms.

C. Use conditional ML estimation to fit a logistic model that contains main effects. What odds ratio dso you obtain for this model? Is there confounding due to SBP and or ECG?

Test
Answers

Chapter 1

True-False Questions:

1. F: any type of independent variable is allowed
2. F: dependent variable must be dichotomous
3. T
4. F: S-shaped
5. T
6. T
7. F: cannot estimate risk using case-control study
8. T
9. F: constant term can be estimated in follow-up study
10. T
11. T
12. F: logit gives log odds, not log odds ratio
13. T
14. F: β_i controls for other variables in the model
15. T
16. F: multiplicative
17. F: $\exp(\beta)$ where β is coefficient of exposure
18. F: OR for effect of SMK is exponential of coefficient of SMK
19. F: OR requires formula involving interaction terms
20. F: OR requires formula that considers coding different from (0, 1)

21. e. $\exp(\beta)$ is not appropriate for **any** X.
22. $P(\mathbf{X}) = 1/(1 + \exp\{-[\alpha + \beta_1(\text{AGE}) + \beta_2(\text{SMK}) + \beta_3(\text{SEX}) + \beta_4(\text{CHOL}) + \beta_5(\text{OCC})]\})$.
23. $P(\mathbf{X}) = 1/(1 + \exp\{-[-4.32 + 0.0274(\text{AGE}) + 0.5859(\text{SMK}) + 1.1523(\text{SEX}) + 0.0087(\text{CHOL}) - 0.5309(\text{OCC})]\})$.
24. $\text{logit } P(\mathbf{X}) = -4.32 + 0.0274(\text{AGE}) + 0.5859(\text{SMK}) + 1.1523(\text{SEX}) + 0.0087(\text{CHOL}) - 0.5309(\text{OCC})$.
25. For a 40-year-old male smoker with CHOL = 200 and OCC = 1, we have
    $\mathbf{X} = (\text{AGE} = 40, \text{SMK} = 1, \text{SEX} = 1, \text{CHOL} = 200, \text{OCC} = 1)$,
    assuming that SMK and SEX are coded as SMK = 1 if smoke, 0 otherwise, and SEX = 1 if male, 0 if female, and
    $P(\mathbf{X}) = 1/(1 + \exp\{-[-4.32 + 0.0274(40) + 0.5859(1) + 1.1523(1) + 0.0087(200) - 0.5309(1)]\})$
    $= 1/\{1 + \exp[-(-0.2764)]\}$
    $= 1/(1 + 1.318)$
    $= 0.431$.

26. For a 40-year-old male *non*smoker with CHOL = 200 and OCC = 1,
    X = (AGE = 40, SMK = 0, SEX = 1, CHOL = 200, OCC = 1)
    and
    $P(\mathbf{X}) = 1/(1 + \exp\{-[-4.32 + 0.0274(40) + 0.5859(0) + 1.1523(1)$
    $\qquad\qquad + 0.0087(200) - 0.5309(1)]\})$
    $\qquad = 1/\{1 + \exp[-(-0.8623)]\}$
    $\qquad = 1/(1 + 2.369)$
    $\qquad = 0.297$

27. The RR is estimated as follows:
    $$\frac{P(\text{AGE} = 40, \text{ SMK} = 1, \text{ SEX} = 1, \text{ CHOL} = 200, \text{ OCC} = 1)}{P(\text{AGE} = 40, \text{ SMK} = 0, \text{ SEX} = 1, \text{ CHOL} = 200, \text{ OCC} = 1)}$$
    $= 0.431/0.297$
    $= 1.45$

    This estimate can be interpreted to say smokers have 1.45 times as high a risk for getting hypertension as nonsmokers, controlling for age, sex, cholesterol level, and occupation.

28. If the study design had been case control or cross sectional, the risk ratio computation of Question 27 would be inappropriate because risk or risk ratio cannot be directly estimated by using a logistic model unless the study design is follow-up. More specifically, the constant term α cannot be estimated from case-control or cross-sectional studies.

29. $\widehat{\text{OR}}$ (SMK controlling for AGE, SEX, CHOL, OCC)
    $= e^{\hat{\beta}}$ where $\hat{\beta} = 0.5859$ is the coefficient of SMK in the fitted model
    $= \exp(0.5859)$
    $= 1.80$

    This estimate indicates that smokers have 1.8 times as high a risk for getting hypertension as nonsmokers, controlling for age, sex, cholesterol, and occupation.

30. The rare disease assumption.

31. The odds ratio is a legitimate measure of association and could be used even if the risk ratio cannot be estimated.

32. $\widehat{\text{OR}}$(OCC controlling for AGE, SEX, SMK, CHOL)
    $= e^{\hat{\beta}}$, where $\hat{\beta} = -0.5309$ is the coefficient of OCC in the fitted model
    $= \exp(-0.5309)$
    $= 0.5881 = 1 / 1.70.$

    This estimate is less than 1 and thus indicates that unemployed persons (OCC = 0) are 1.70 times more likely to develop hypertension than are employed persons (OCC = 1).

33. Characteristic 1: the model contains only main effect variables
    Characteristic 2: OCC is a (0, 1) variable.

34. The formula $\exp(\beta_i)$ is inappropriate for estimating the effect of AGE controlling for the other four variables because AGE is being treated as a continuous variable in the model, whereas the formula is appropriate for (0, 1) variables only.

Chapter 2

True-False Questions:

1. F: OR = $\exp(\psi)$
2. F: risk = $1/[1 + \exp(-\alpha)]$
3. T
4. T
5. T
6. T
7. T
8. F: OR = $\exp(\beta + 5\delta)$
9. F: the number of dummy variables should be 19
10. F: OR = $\exp(\beta + \delta_1 \text{OBS} + \delta_2 \text{PAR})$

11. The model in logit form is given as follows:
    logit $P(\mathbf{X}) = \alpha + \beta\text{CON} + \gamma_1\text{PAR} + \gamma_2\text{NP} + \gamma_3\text{ASCM} + \delta_1\text{CON} \times \text{PAR} + \delta_2\text{CON} \times \text{NP} + \delta_3\text{CON} \times \text{ASCM}$.
12. The odds ratio expression is given by
    $\exp(\beta + \delta_1\text{PAR} + \delta_2\text{NP} + \delta_3\text{ASCM})$.
13. The model for the matched pairs case-control design is given by
    $$\text{logit } P(\mathbf{X}) = \alpha + \beta\text{CON} + \sum_{i=1}^{199}\gamma_i V_i + \gamma_{200}\text{PAR} + \delta\text{CON} \times \text{PAR},$$
    where the V_i are dummy variables which indicate the matching strata.
14. The risk for an exposed person in the first matched pair with PAR = 1 is
    $$R = \frac{1}{1 + \exp\left[-\left(\alpha + \beta + \gamma_1 + \gamma_{200} + \delta\text{PAR}\right)\right]}.$$
    The odds ratio expression for the matched pairs case-control model is $\exp(\beta + \delta\text{PAR})$

Chapter 3

1. a. ROR = $\exp(\beta)$
   b. ROR = $\exp(5\beta)$
   c. ROR = $\exp(2\beta)$
   d. All three estimated odds ratios should have the same value.
   e. The β in part b is one-fifth the β in part a; the β in part c is one-half the β in part a.

2. a. $ROR = \exp(\beta + \delta_1 AGE + \delta_2 CHL)$
    b. $ROR = \exp(5\beta + 5\delta_1 AGE + 5\delta_2 CHL)$
    c. $ROR = \exp(2\beta + 2\delta_1 AGE + 2\delta_2 CHL)$
    d. For a given specification of AGE and CHL, all three estimated odds ratio should have the same value.
    e. The β in part b is one-fifth the β in part a; the β in part c is one-half the β in part a. The same relationships hold for the three δ_1's and the three δ_2's.

3. a. $ROR = \exp(50\beta + 50\delta_1 AGE + 50\delta_2 SEX)$
    b. $ROR = \exp(10\beta + 10\delta_1 AGE + 10\delta_2 SEX)$
    c. $ROR = \exp(10\beta + 10\delta_1 AGE + 10\delta_2 SEX)$
    d. For a given specification of AGE and SEX, the odds ratios in parts b and c should have the same value.

4. a. $logit P(\mathbf{X}) = \alpha + \beta_1 S_1 + \beta_2 S_2 + \gamma_1 AGE + \gamma_2 SEX$, where S_1 and S_2 are dummy variables which distinguish between the three SSU groupings, e.g., $S_1 = 1$ if low, 0 otherwise and $S_2 = 1$ if medium, 0 otherwise.
    b. Using the above dummy variables, the odds ratio is given by $ROR = \exp(-\beta_1)$, where $\mathbf{X}^* = (0, 0, AGE, SEX)$ and $\mathbf{X}^{**} = (1, 0, AGE, SEX)$.
    c. $logit P(\mathbf{X}) = \alpha + \beta_1 S_1 + \beta_2 S_2 + \gamma_1 AGE + \gamma_2 SEX + \delta_1 (S_1 \times AGE) + \delta_2 (S_1 \times SEX) + \delta_3 (S_2 \times AGE) + \delta_4 (S_2 \times SEX)$
    d. $ROR = \exp(-\beta_1 - \delta_1 AGE - \delta_2 SEX)$

5. a. $ROR = \exp(10\beta_3)$
    b. $ROR = \exp(195\beta_1 + 10\beta_3)$

6. a. $ROR = \exp(10\beta_3 + 10\delta_{31} AGE + 10\delta_{32} RACE)$
    b. $ROR = \exp(195\beta_1 + 10\beta_3 + 195\delta_{11} AGE + 195\delta_{12} RACE + 10\delta_{31} AGE + 10\delta_{32} RACE)$

Chapter 4

True-False Questions:

1. T
2. T
3. F: unconditional
4. T
5. F: the model contains a large number of parameters
6. T
7. T
8. F: α is not estimated in conditional ML programs
9. T
10. T
11. F: the variance–covariance matrix gives variances and covariances for regression coefficients, not variables.

12. T

13. Because matching has been used, the method of estimation should be **conditional** ML estimation.

14. The variables AGE and SOCIOECONOMIC STATUS do not appear in the printout because these variables have been matched on, and the corresponding parameters are nuisance parameters that are not estimated using a conditional ML program.

15. The OR is computed as e to the power 0.39447, which equals 1.48. This is the odds ratio for the effect of pill use adjusted for the four other variables in the model. This odds ratio says that pill users are 1.48 times as likely as nonusers to get cervical cancer after adjusting for the four other variables.

16. The OR given by e to -0.24411, which is 0.783, is the odds ratio for the effect of vitamin C use adjusted for the effects of the other four variables in the model. This odds ratio says that vitamin C is somewhat protective for developing cervical cancer. In particular, since 1/0.78 equals 1.28, this OR says that vitamin C **nonusers** are 1.28 times more likely to develop cervical cancer than **users,** adjusted for the other variables.

17. Alternative null hypotheses:
    1. The OR for the effect of VITC adjusted for the other four variables equals 1.
    2. The coefficient of the VITC variable in the fitted logistic model equals 0.

18. The 95% CI for the effect of VITC adjusted for the other four variables is given by the limits 0.5924 and 1.0359.

19. The Z statistic is given by $Z = -0.24411/0.14254 = 1.71$.

20. The value of MAX LOGLIKELIHOOD is the logarithm of the maximized likelihood obtained for the fitted logistic model. This value is used as part of a likelihood ratio test statistic involving this model.

Chapter 5

1. Conditional ML estimation is the appropriate method of estimation because the study involves matching.

2. Age and socioeconomic status are missing from the printout because they are matching variables and have been accounted for in the model by nuisance parameters which are not estimated by the conditional estimation method.

3. H_0: $\beta_{SMK} = 0$ in the no interaction model (model I), or alternatively, H_0: OR=1, where OR denotes the odds ratio for the effect of SMK on cervical cancer status, adjusted for the other variables (NS and AS) in model I;

   test statistic: Wald statistic $Z = \dfrac{\hat{\beta}_{SMK}}{S_{\hat{\beta}_{SMK}}}$, which is approximately normal (0, 1) under H_0, or alternatively,

Z^2 is approximately chi square with one degree of freedom under H_0; test computation: $Z = \dfrac{1.4361}{0.3167} = 4.53$; alternatively, $Z^2 = 20.56$; the one-tailed P-value is 0.0000/2 = 0.0000, which is highly significant.

4. The point estimate of the odds ratio for the effect of SMK on cervical cancer status adjusted for the other variables in model I is given by $e^{1.4361} = 4.20$.

   The 95% interval estimate for the above odds ratio is given by
   $$\exp\left[\hat{\beta}_{SMK} \pm 1.96\sqrt{\widehat{\mathrm{Var}}(\hat{\beta}_{SMK})}\right] = \exp(1.4361 \pm 1.96 \times 0.3617)$$
   $$= \left(e^{0.7272},\, e^{2.1450}\right) = (2.07,\ 8.54).$$

5. Null hypothesis for the likelihood ratio test for the effect of SMK × NS: H_0: $\beta_{SMK \times NS} = 0$ where $\beta_{SMK \times NS}$ is the coefficient of SMK × NS in model II;

   Likelihood ratio statistic: LR $= -2 \ln \hat{L}_I - (-2 \ln \hat{L}_{II})$ where $\hat{L}_I$ and $\hat{L}_{II}$ are the maximized likelihood functions for models I and II, respectively. This statistic has approximately a chi-square distribution with one degree of freedom under the null hypothesis.

   Test computation: LR = 174.97 − 171.46 = 3.51. The P-value is less than 0.10 but greater than 0.05, which gives borderline significance because we would reject the null hypothesis at the 10% level but not at the 5% level. Thus, we conclude that the effect of the interaction of NS with SMK is of borderline significance.

6. Null hypothesis for the Wald test for the effect of SMK × NS is the same as that for the likelihood ratio test: H_0: $\beta_{SMK \times NS} = 0$ where $\beta_{SMK \times NS}$ is the coefficient of SMK × NS in model II;

   Wald statistic: $Z = \dfrac{\hat{\beta}_{SMK \times NS}}{s_{\hat{\beta}_{SMK \times NS}}}$, which is approximately normal (0, 1) under H_0, or alternatively,

   Z^2 is approximately chi square with one degree of freedom under H_0; test computation: $Z = \dfrac{-1.1128}{0.5997} = -1.856$; alternatively, $Z^2 = 3.44$; the P-value for the Wald test is 0.0635, which gives borderline significance.

   The LR statistic is 3.51, which is approximately equal to the square of the Wald statistic; therefore, both statistics give the same conclusion of borderline significance for the effect of the interaction term.

7. The formula for the estimated odds ratio is given by
   $$\widehat{OR}_{adj} = \exp\left(\beta_{SMK} + \delta_{SMK \times NS}NS\right) = \exp(1.9381 - 1.1128\ NS)$$
   where the coefficients come from model II and the confounding effects of NS and AS are controlled.

8. Using the adjusted odds ratio formula given in Question 7, the estimated odds ratio values for NS = 1 and NS = 0 are

NS = 1:   exp[1.9381 − 1.1128(1)] = exp(0.8253) = 2.28;
NS = 0:   exp[1.9381 − 1.1128(0)] = exp(1.9381) = 6.95

9. Formula for the 95% confidence interval for the adjusted odds ratio when NS = 1:

$$\exp\left[\hat{l} \pm 1.96\sqrt{\widehat{\text{var}}(\hat{l})}\right], \text{ where } \hat{l} = \hat{\beta}_{SMK} + \hat{\delta}_{SMK \times NS}(1) = \hat{\beta}_{SMK} + \hat{\delta}_{SMK \times NS}$$

and

$$\widehat{\text{var}}(\hat{l}) = \widehat{\text{var}}(\hat{\beta}_{SMK}) + (1)^2 \widehat{\text{var}}(\hat{\delta}_{SMK \times NS}) + 2(1)\widehat{\text{cov}}(\hat{\beta}_{SMK}, \hat{\delta}_{SMK \times NS}),$$

where $\widehat{\text{var}}(\hat{\beta}_{SMK})$, $\widehat{\text{var}}(\hat{\delta}_{SMK \times NS})$, and $\widehat{\text{cov}}(\hat{\beta}_{SMK}, \hat{\delta}_{SMK \times NS})$ are obtained from the printout of the variance–covariance matrix.

10. $\hat{l} = \hat{\beta}_{SMK} + \hat{\delta}_{SMK \times NS} = 1.9381 + (-1.1128) = 0.8253$

$$\widehat{\text{var}}(\hat{l}) = 0.1859 + (1)^2 (0.3596) + 2(1)(-0.1746) = 0.1859 + 0.3596 - 0.3492 = 0.1963.$$

The 95% confidence interval for the adjusted odds ratio is given by

$$\exp\left[\hat{l} \pm 1.96\sqrt{\widehat{\text{Var}}(\hat{l})}\right] = \exp(0.8253 \pm 1.96\sqrt{0.1963}) = \exp(0.8253 \pm 1.96 \times 0.4430)$$

$$= (e^{-0.0430}, e^{1.6936}) = (0.96, 5.44).$$

11. Model II is more appropriate than model I if the test for the effect of interaction is viewed as significant. Otherwise, model I is more appropriate than model II. The decision here is debatable because the test result is of borderline significance.

Chapter 6

True-False Questions:

1. F: one stage is variable specification
2. T
3. T
4. F: no statistical test for confounding
5. F: validity is preferred to precision
6. F: for initial model, V's chosen a priori
7. T
8. T
9. F: model needs $E \times B$ also
10. F: list needs to include $A \times B$

11. The given model is hierarchically well formulated because for each variable in the model, every lower-order component of that variable is contained in the model. For example, if we consider the variable SMK × NS × AS, then the lower-order components are SMK, NS, AS, SMK × NS, SMK × AS, and NS × AS; all these lower-order components are contained in the model.

12. A test for the term SMK × NS × AS is not dependent on the coding of SMK because the model is hierarchically well formulated and SMK × NS × AS is the highest-order term in the model.

13. A test for the terms SMK × NS is dependent on the coding because this variable is a lower-order term in the model, even though the model is hierarchically well formulated.

14. In using a hierarchical backward elimination procedure, first test for significance of the highest-order term SMK × NS × AS, then test for significance of lower-order interactions SMK × NS and SMK × AS, and finally assess confounding for *V* variables in the model. Based on the Hierarchy Principle, any two-factor product terms and *V* terms which are lower-order components of higher-order product terms found significant are not eligible for deletion from the model.

15. If SMK × NS × AS is significant, then SMK × NS and SMK × AS are interaction terms that must remain in any further model considered. The *V* variables that must remain in further models are NS, AS, NS × AS, and, of course, the exposure variable SMK. Also the *V** variables must remain in all further models because these variables reflect the matching that has been done.

16. The model after interaction assessment is the same as the initial model. No potential confounders are eligible to be dropped from the model because NS, AS, and NS × AS are lower components of SMK × NS × AS and because the *V** variables are matching variables.

Chapter 7

1. The interaction terms are SMK × NS, SMK × AS, and SMK × NS × AS. The product term NS × AS is a *V* term, not an interaction term, because SMK is not one of its components.

2. Using a hierarchically backward elimination strategy, one would first test for significance of the highest-order interaction term, namely, SMK × NS × AS. Following this test, the next step is to evaluate the significance of two-factor product terms, although these terms might not be eligible for deletion if the test for SMK × NS × AS is significant. Finally, without doing statistical testing, the *V* variables need to be assessed for confounding and precision.

3. If SMK × NS is the only interaction found significant, then the model remaining after interaction assessment contains the V^* terms, SMK, NS, AS, NS × AS, and SMK × NS. The variable NS cannot be deleted from any further model considered because it is a lower-order component of the significant interaction term SMK × NS. Also, the V^* terms cannot be deleted because these terms reflect the matching that has been done.

4. The odds ratio expression is given by $\exp(\beta + \delta_1 NS)$.

5. The odds ratio expression for the model that does not contain NS × AS has exactly the same form as the expression in Question 4. However, the coefficients β and δ_1 may be different from the Question 4 expression because the two models involved are different.

6. Drop NS × AS from the model and see if the estimated odds ratio changes from the gold standard model remaining after interaction assessment. If the odds ratio changes, then NS × AS cannot be dropped and is considered a confounder. If the odds ratio does not change, then NS × AS is not a confounder. However, it may still need to be controlled for precision reasons. To assess precision, one should compare confidence intervals for the gold standard odds ratio and the odds ratio for the model that drops NS × AS. If the latter confidence interval is meaningfully narrower, then precision is gained by dropping NS × AS, so that this variable should, therefore, be dropped. Otherwise, one should control for NS × AS because no meaningful gain in precision is obtained by dropping this variable. Note that in assessing both confounding and precision, tables of odds ratios and confidence intervals obtained by specifying values of NS need to be compared because the odds ratio expression involves an effect modifier.

7. If NS × AS is dropped, the only V variable eligible to be dropped is AS. As in the answer to Question 6, confounding of AS is assessed by comparing odds ratio tables for the gold standard model and reduced model obtained by dropping AS. The same odds ratio expression as given in Question 5 applies here, where, again, the coefficients for the reduced model (without AS and NS × AS) may be different from the coefficient for the gold standard model. Similarly, precision is assessed similarly to that in Question 6 by comparing tables of confidence intervals for the gold standard model and the reduced model.

8. The odds ratio expression is given by $\exp(1.9381 - 1.1128 NS)$. A table of odds ratios for different values of NS can be obtained from this expression and the results interpreted. Also, using the estimated variance–covariance matrix (not provided here), a table of confidence intervals (CIs) can be calculated and interpreted in conjunction with corresponding odds ratio estimates. Finally, the CIs can be used to carry out two-tailed tests of significance for the effect of SMK at different levels of NS.

Chapter 8

1. T
2. F: information may be lost from matching: sample size may be reduced by not including eligible controls
3. T
4. T
5. T
6. McNemar's chi square: $(X - Y)^2/(X + Y) = (125 - 121)^2/(125 + 121) = 16/246 = 0.065$, which is highly nonsignificant. The MOR equals $X/Y = 125/121 = 1.033$. The conclusion from this data is that there is no meaningful or significant effect of exposure (Vietnam veteran status) on the outcome (genetic anomalies of offspring).

7.
$$\text{logit } P(\mathbf{X}) = \alpha + \beta E + \sum_{i=1}^{8501} \gamma_{1i} V_{1i},$$

   where the V_{1i} denote 8501 dummy variables used to indicate the 8502 matched pairs.

8. The Wald statistic is computed as $Z = 0.032/0.128 = 0.25$. The square of this Z is 0.0625, which is very close to the McNemar chi square of 0.065, and is highly nonsignificant.

9. The odds ratio from the printout is 1.033, which is identical to the odds ratio obtained using the formula X/Y.

10. The confidence interval given in the printout is computed using the formula

$$\exp\left[\hat{\beta} \pm 1.96\sqrt{\widehat{\text{var}}(\hat{\beta})}\right],$$

   where the estimated coefficient $\hat{\beta}$ is 0.032 and the square root of the estimated variance, i.e., $\sqrt{\widehat{\text{var}}(\hat{\beta})}$, is 0.128.

11. Conditional ML estimation was used to fit the model because the study design involved matching; the number of parameters in the model, including dummy variables for matching, is large relative to the number of observations in the study.

12. The variables age and socioeconomic status are missing from the printout because they are matching variables and have been accounted for in the model by nuisance parameters (corresponding to dummy variables) which are not estimated by the conditional ML estimation method.

13. $\text{logit } P(\mathbf{X}) = \alpha + \beta \text{SMK} + \sum_{i} \gamma_{1i} V_{1i} + \gamma_{21} \text{NS} + \gamma_{22} \text{AS} + \delta \text{SMK} \times \text{NS},$

   where V_{1i} denote dummy variables (the number not specified) used to indicate matching strata.

14. The squared Wald statistic (a chi-square statistic) for the product term SMK $\times$ NS is computed to be 3.44, which has a *P*-value of 0.0635. This is not significant at the 5% level but is significant at the 10% level. This is not significant using the traditional 5% level, but it is close enough to 5% to suggest possible interaction. Information for computing the likelihood ratio test is not provided here, although the likelihood ratio test would be more appropriate to use in this situation. (Note, however, that the LR statistic is 3.51, which is also not significant at the 0.05 level, but significant at the 0.10 level.)

15. The formula for the estimated odds ratio is given by
$$\widehat{OR}_{adj} = \exp(\hat{\beta}_{SMK} + \hat{\delta}_{SMK \times NS} NS) = \exp(1.9381 - 1.1128NS),$$
where the coefficients come from the printout and the confounding effects of NS and AS are controlled.

16. Using the adjusted odds ratio formula given in Question 8, the estimated odds ratio values for NS = 1 and NS = 0 are
NS = 1: $\exp[1.9381 - 1.1128(1)] = \exp[0.8253] = 2.28$,
NS = 0: $\exp[1.9381 - 1.1128(0)] = \exp[1.9381] = 6.95$.

17. When NS = 1, the point estimate of 2.28 shows a moderate effect of smoking, i.e., the risk for smokers is about 2.3 times the risk for non-smokers. However, the confidence interval of (0.96, 5.44) is very wide and contains the null value of 1. This indicates that the point estimate is strongly nonsignificant and is highly unreliable. Conclude that, for NS = 1, there is no significant evidence to indicate that smoking is related to cervical cancer.

18. Using the printout for the no interaction model, the estimated odds ratio adjusted for the other variables in model 1 is given by $e^{1.4361} = 4.20$.

The Wald statistic $Z = \dfrac{\hat{\beta}_{SMK}}{S_{\hat{\beta}_{SMK}}}$ is approximately normal (0, 1) under H_0, or alternatively,

Z^2 is approximately chi square with one degree of freedom under H_0;

test computation: $Z = \dfrac{1.4361}{0.3167} = 4.53$; alternatively, $Z^2 = 20.56$;

the one-tailed *P*-value is 0.0000/2 = 0.0000, which is highly significant. The 95% interval estimate for the above odds ratio is given by

$$\exp\left[\hat{\beta}_{SMK} \pm 1.96\sqrt{\widehat{var}(\hat{\beta}_{SMK})}\right] = \exp(1.4361 \pm 1.96 \times .3617)$$
$$= \left(e^{0.7272}, e^{2.1450}\right) = (2.07, 8.54).$$

The above statistical information gives a meaningfully significant point estimate of 4.20, which is statistically significant. The confidence interval is quite wide, ranging between 2 and 8, but the lower limit of 2 suggests that there is real effect of smoking in this data set. All of these results depend on the no interaction model being the correct model, however.

Bibliography

Bishop, Y.M.M., Fienberg, S.E., and Holland, P.W., *Discrete Multivariate Analysis: Theory and Practice*, MIT Press, Cambridge, Mass., 1975.

Breslow, N.E., and Day, N.E., *Statistical Methods in Cancer Research, Vol. 1: The Analysis of Case-Control Studies*, IARC Scientific Publications No. 32, Lyon, France, 1981.

*Brock, K.E., Berry, G., Mock, P.A., MacLennan, R., Truswell, A.S., Brinton, L.A., "Nutrients in Diet and Plasma and Risk of In Situ Cervical Cancer," *J. Nat. Cancer Inst.*, Vol. 80(8):580–585, 1988.

*Donovan et al, "Agent Orange," *Medical Journal of Australia*, Vol. 145, 1984.

Kleinbaum, D.G., Kupper L.L., and Chambless, L.E., "Logistic regression analysis of epidemiologic data: theory and practice," *Commun. Stat.*, Vol. 11(5):485–547, 1982.

Kleinbaum, D.G., Kupper L.L., and Morgenstern, H., *Epidemiologic Research: Principles and Quantitative Methods*, Van Nostrand Reinhold Publishers, New York, 1982.

Kleinbaum, D.G., Kupper L.L., and Muller, K.A., *Applied Regression Analysis and Other Multivariable Methods, Second Edition*, Duxbury Press, Boston, 1987.

*McLaws, M., Irwig, L.M., Mock, P., Berry, G. and Gold, J., "Predictors of surgical wound infection in Australia: a national study," *Medical Journal of Australia*, Vol. 149:591–595, 1988.

Prentice, R.L. and Pyke, R., "Logistic disease incidence models and case-control studies," *Biometrics*, Vol. 32(3):599–606, 1976.

*Sources for practice exercises or test questions presented at the end of several chapters.

Index